CARBON COLONIALISM

Manchester University Press

CARBON COLONIALISM

How rich countries export climate breakdown

LAURIE PARSONS

Manchester University Press

Copyright © Laurie Parsons 2023

The right of Laurie Parsons to be identified as the author of this work has been asserted in accordance with the Copyright, Designs and Patents Act 1988.

Published by Manchester University Press
Oxford Road, Manchester M13 9PL

www.manchesteruniversitypress.co.uk

British Library Cataloguing-in-Publication Data
A catalogue record for this book is available from the British Library

ISBN 978 1 5261 6918 1 hardback

First published 2023

The publisher has no responsibility for the persistence or accuracy of URLs for any external or third-party internet websites referred to in this book, and does not guarantee that any content on such websites is, or will remain, accurate or appropriate.

Typeset
by Cheshire Typesetting Ltd, Cuddington, Cheshire
Printed in Great Britain
by TJ Books Ltd, Padstow, Cornwall

Contents

1 Moving forwards, or dumping sideways? The myth
of a sustainable future *page* 1

Part I Greenwashing the global factory

2 Founding the global factory: the first five hundred
years 25
3 Consumer power in the global factory: a lucrative
illusion 48
4 Carbon colonialism: hidden emissions in the global
periphery 76

Part II Manufacturing disaster in the global factory

5 Climate precarity: how global inequality shapes
environmental vulnerability 101
6 Money talks: who gets to speak for the environment
and how 124

Contents

7 Wolves in sheep's clothing: how corporate logic
co-opts climate action · · · 152

8 Six myths that fuel carbon colonialism – and how to
think differently · · · 176

Notes · · · 208
Index · · · 228

1

Moving forwards, or dumping sideways? The myth of a sustainable future

In early 2018, I found myself walking up the side of a landfill site on the outskirts of the Cambodian capital, Phnom Penh. The air around me clung heavy and immobile in the sweltering midday sun, but the dump itself was alive with activity. A dozen or so cows chewed lazily at scraps of fabric, interspersed by about double that number of waste pickers collecting scraps of fabric into burlap sacks. On occasion, a zippy mechanical digger would mount the side of the rubbish heap, delivering a fresh load of textile waste. Scraps of clothing mingled with bundles of tangled thread, or plastic bags full of cable ties. On occasion, an entire bag of clothing labels for a well-known brand would be deposited on the ever-growing mound, but they were visible almost everywhere anyway. Marks & Spencer, George, Pull & Bear, Walmart, Gap, the dump was a who's who of major clothing labels: every one of them proudly declaring their commitment to sustainability and elimination of landfill from their supply chain.

Yet the labels themselves were no surprise. After all, I hadn't tracked them here directly from the factory. I hadn't, in fact, been investigating clothing at all. Instead, I had begun work a few months previously on a project exploring debt bondage in the Cambodian brick sector, a brutal industry in which heavily indebted farmers indenture themselves for years or even decades to pay off existing loans. It was only on visiting these kilns that I realised how many used fabric scraps as fuel to fire their bricks. The dirt floor of the kilns, strewn with clothing labels for major global brands, revealed a dirty and little-known secret. Garment factories were selling waste to companies who brought it to landfill, some of which would then end up as toxic smoke swirling around debt-bonded workers and child labourers trapped in an industry without hope.

It's a chilling thought: the idea that innocuous everyday goods could be linked to human suffering and environmental destruction on such a scale. It feels like an aberration – something that can and should be stamped out – which makes it in some ways a bad example. In reality, almost everything we buy involves exploiting the environment and the people who depend on it to a greater or lesser extent. Almost everything we buy makes a contribution to climate breakdown, through emissions, local environmental degradation, or most commonly both. Yet in a world where greenwashing is so commonplace that almost every product proclaims environmental benefits, it tends not to be seen that way. In fact, it tends not to be seen at all.

So, what do we see? What comes into your mind when you think of environmental breakdown? If you're living in the rich world it may well be one of the following: a polar bear stranded in melting tundra, forest fires blazing in the Amazon, thousands of miles of sea filled with plastic, drought-hit communities sweltering in extreme temperatures. In a minority of cases, your example might be taken from the West: hurricane Katrina,

The myth of a sustainable future

or the floods that wreaked havoc on German villages in 2021. What your example is unlikely to include is your own town, your own neighbourhood, your own street – because whether we see it as more or less distant, the climate crisis is nonetheless always a step removed. It's not you. It never is.

For citizens of the wealthy global North this is far from illogical. Europe and the US have thus far suffered the impacts of climate change to a far lesser degree than their less wealthy global Southern counterparts. Crucially, they also have far greater resources to mitigate these impacts. The Netherlands and Bangladesh are geographically similar in terms of their natural vulnerability to climate change, yet flooding has not been a serious issue in the former since the national dam-building programme of the mid-twentieth century. When it comes to climate change impacts, money matters. The North has it, the South doesn't – and the geography of climate risk reflects that.

The same thing applies to the local environment. The rich world has made grand strides towards repairing the damage inflicted by industrialisation. London's air pollution, once so notoriously unhealthy that each 'pea souper' fog would result in dozens of deaths, has, in common with most of Europe's major cities, improved markedly in recent decades. Fish have returned to the great rivers of the continent, from the Thames to the Danube to the Rhine. The rich world has, it seems, *progressed* past the stage of dirty industry. Our economies have *evolved* out of our dependency on choking fumes and toxic chemicals, learning new, cleaner ways to replace the old.

That other parts of the world have not followed the same path is no challenge to this notion. The choking air pollution of Delhi and Beijing, the Citarum river of West Java that has been so polluted by the five hundred factories that line its banks that fishing communities have given up trying to find fish and turned instead to collecting the far more plentiful plastic waste to sell to recyclers. Cases like these are fodder to the rich world's belief in environmental progress: bolstering the notion that 'once we were like this, but soon they will follow in our footsteps'. There is even a scientific basis for this reasoning: the Environmental Kuznets Curve that, when mapping pollution to GDP, appears to show a rise and subsequent decline in the course of a country's economic development.

It's an appealing argument. And, implicitly or explicitly, it has become firmly entrenched into the discourse on climate change. Carbon emissions and pollution are a phase that we all pass through, meaning that the ability – and crucially the money – to avoid the ratcheting risks of climate change is something we have earned, and others too will earn as each nation continues inexorably along its separate curve. For wealthy countries, this narrative is accepted because it is comfortable, because it provides a logical and moral explanation of the relative safety and health of the rich world. But what if it wasn't true? What if one place was devastated *because* the other was clean? What if one place was at risk *because* the other was safe?

It's an idea that's not totally unfamiliar. The idea of climate justice, brought to the mainstream by groups such as Fridays for Future and Extinction Rebellion, emphasises the historical responsibility of rich countries for the current climate change the world is facing. It also highlights the difference in vulnerability faced by the highest and lowest emitters, comparing climate change to 'second-hand smoke'[1] for its ability to harm those with little to do with producing it. This inequality in the generation and impacts of carbon emissions is vitally important to understanding how and why climate change manifests, but it

The myth of a sustainable future

is not the whole story. The vulnerability of those populations is not simply a coincidence. Just as carbon emissions are not acts of God, neither is exposure to the results of those emissions. Both are rooted in the same economic logic.

Take sea-level rises for example. This is known to be one of the most destructive impacts of anthropogenic climate change. Under a medium warming scenario, sea levels are expected to rise by 0.6–1.1m by the middle of the century,[2] leaving up to 1 billion square kilometres of low-lying coastal regions either underwater or within a metre of being so.[3] It is a compelling vision of the future, as much for its simplicity as for its apparent inevitability. We know that temperatures will rise. We know that higher temperatures will melt glaciers, which will raise sea levels. Those low-lying regions are therefore as good as gone already, their populations displaced to who knows where.

The problem is that it isn't that simple. Human societies are powerful and resourceful and – on a local scale at least – possess ample ability to manipulate nature. It is often pointed out that much of the Netherlands was underwater prior to a national dyke-building exercise that began in the eleventh century and continued for hundreds of years. Somewhat less appreciated is the everyday miracle of London's Thames Barrier, that rises and falls to keep the UK capital safe from the regular inundation it would now otherwise be facing. Venice has similar systems in place, and even inland cities have evolved protection from unexpected rises in water levels. Following a highly destructive flood in 1910, for example, the French government began work on a series of reservoirs called Les Grands Lacs de Seine (Great Lakes of the Seine) in order to mitigate flood pressures on Paris's arterial river and ensure that misery on the scale of the Great Flood would never happen again.

Carbon colonialism

These examples show that vulnerability to the hazards emerging from climate change is by no means inevitable. It is a choice, or, more pertinently, a function of wealth and its absence. Hazards like storms, floods, and rising seas are increasing in number and intensity under climate change, but they are nothing fundamentally new. There are tried and tested ways of mitigating them, but they are expensive. Jakarta does not have the resources to combat rising seas that London or Venice does. Bangladesh has nowhere near the capital necessary to construct flood defences on a similar scale to those that protect the 59 per cent of the Netherlands that is at or below sea level.

You can't, in other words, remove money from the geography of disaster risk. Haiti, Myanmar, Bangladesh, Pakistan, to name but a few, are faced with landslides, droughts, floods, and extreme heat, which they know will worsen in the coming years. For millions of people, this means disrupted farming and food shortages. But it doesn't *have* to mean this. It means this because of a system in which the environmental cost of wealth generation is paid in places far from where that wealth is accumulated. This, as I call it in this book, is carbon colonialism: the latest incarnation of an age-old system in which natural resources continue to be extracted, exported, and profited from far from the people they used to belong to. It is, in many ways, an old story, but what is new is the hidden cost of that extraction: the carbon bill footed in inverse relation to the resource feast.

And there is a further dimension to this also. Environmental risk is not only a question of natural hazards, but also of the local conditions it meets. The impact of a flood will be far worse if that flood water is full of human waste and toxic chemicals, as is the case in so many flood-threatened areas of the global South, not least our previous example of Jakarta, where

The myth of a sustainable future

a public health catastrophe is threatened each time rising sea levels, heavy rains, and the world's most polluted waterways converge. Similarly, the flooding and droughts brought about by unpredictable rains in Bangladesh would be far easier to cope with without the smoke, heat, and millions of cubic metres of farmland harvested by the brick industry. Climate change is a global problem, but local economic and industrial factors play a major role in shaping its harms.

Much of the remainder of this book will be devoted to exploring this global economy of environmental risk, its origins, details, and consequences. Yet at the heart of its investigation is a question: if global production plays such a big role in shaping vulnerability, why do we still call disasters natural? As this book argues, the hidden nature of environmental degradation is a key part of the story, rooted as it is in core beliefs about the nature of economic growth, and technological and social progress, that are rapidly becoming unstuck.

Evolving towards sustainability?

On my first trip to visit Cambodia's brick kilns in 2017, I watched an old man hurl bag after bag of fabric into a roaring furnace. Thick, black smoke swirled from the low chimney, whilst searching fingers of molten plastic felt their way from beneath the flames. A young boy, the son of a worker and an occasional worker himself, coughed and walked the twenty feet to his aluminium home, picking his way through labels that a few weeks ago I myself had been picking through on the rack of a London clothes shop.

It was a moment of cognitive dissonance. Yet scenes like this, combining emissions, environmental degradation, global production, and poverty, are repeated millions of times every day around the world. The problem is that from the perspective of corporate sustainability they don't exist. Global production as it is displayed to the world is simple, clean, and decarbonising. The reality is far from it: an unregulated and unseen world of carbon-intensive production and local environmental destruction. This hidden world of global production is the new frontier of the fight against climate breakdown. Not only does it undermine our ability to tackle global emissions, but smaller-scale impacts too are hidden amidst the complex logistics of our global production networks.

It is hidden because of the vast changes that have been underway in our societies and economies in the last half century. Since the 1970s, when the world's leading economies were also the largest manufacturers of everyday goods like t-shirts and toasters for their domestic markets,[4] the majority of what consumers in the rich world use in their everyday lives is now produced overseas. Economies around the world have expanded and deepened their overseas activities. Supply chains have become more complex and international. It is rare now for a garment you might pick from a hanger in a shop to originate entirely from one country. Most items of clothing are processed in multiple fields, factories, and nations.

Arrangements like these lower costs and generate efficiencies, yet they also produce obscurity. The longer a supply chain, the harder it is to keep track of, the more intermediaries are involved, the less the oversight. More important still, in crossing borders, they cross not only logistical checkpoints, but legal and political worlds. Let's say you bring home a toaster marked 'Made in Vietnam' from a high street retailer. What environmental standards does it adhere to? The assembly process

The myth of a sustainable future

might comply with Vietnamese ones, but the 157 components in a standard toaster[5] have known Vietnam only fleetingly. A toaster includes steel, zinc, plastic, copper, and nickel, but let's just take the last of these. Where does the nickel in your toaster come from? Certainly not Vietnam, which has only one sporadically operating nickel mine. We know that Indonesia produces about a third of global nickel at the moment, so there is a decent chance it comes from there, but the truth is we don't know.

The scary part comes when we begin to ask who does know. Not the owners of the shop, or even the company. Neither have any direct oversight over the supply chain of most products. Not the assembling factory in Vietnam: they know where they purchase the components from, but they don't know where the raw materials in them originate. Even the people who make the components may not be sure where those raw materials originate because in most cases they will have purchased them from an intermediary supplier.

And this is before we consider the question of legality. Although almost every country in the world has ratified environmental and labour standards of some description, enforcement of these standards differs hugely. Many countries in the global South lack the capacity to rigorously monitor industrial activities, and even where they do, corruption is not uncommon. This may mean dumping wastewater in the local river – which, it should be noted, European factories have done for centuries – or it may mean the production of airborne pollutants that would not be tolerated in the countries that purchase the goods those factories make. Crucially, these sorts of environmental inequalities rarely take the form of major infringements explicitly recorded in supply chains outlined

Carbon colonialism

to consumers. It is never noted on a label that the factory that makes your clothing pumps wastewater filled with glue and dye into a sewer emerging in a nearby lake. These questions are simply never asked. The factories claim to comply with national standards and that is usually deemed sufficient to probe no further.

On one level, this seems perfectly reasonable: after all, it is more or less how production has always worked. Yet in the recent, globalised era, the role of distance and borders is accentuated. In a traditional industrial manufacturing context, a company – let's say a shoe company, British Shoes Inc. – may own a number of factories in a given area, let's say Lancashire. If one of the factories making shoes for this company began to pump industrial waste into the local river, killing fish and leading to illness in the local community, then the members of that community would know what to do. They would take their complaint to local government, who would appeal to the factory to stop the practice. If this was not successful, the complaint would be passed on to national government and environmental agencies that govern not only the factory but also the company itself, with sanctions applied if necessary.

This is, of course, an idealised scenario. Much domestic industry continues to do substantial harm to the environments in which it is situated. Yet it bears outlining because of the contrast with how a similar problem would be dealt with in a globalised supply chain. Let's take an equivalent scenario: shoes being produced for UK consumers, not now in Lancashire, but in the province of Kampong Speu in Southern Cambodia. If local people wish to complain about glue in their river, then they can similarly lodge a complaint with local government. Yet the reality is that the owner of a factory in such a context

The myth of a sustainable future

is a powerful person. They may well find a way to hide the practice, or simply block any such complaint, as happens on innumerable such occasions.

The problem arises when local people might wish to escalate their complaint. Unlike in the domestic example, the factory is not owned by the company that sells the goods. Sanctions may be taken against an individual factory, but there is no way to ensure the largest economic actor – the brand – enforces these standards more broadly. Indeed, based as they usually are in a jurisdiction far from the production process, government authorities in producer countries have very little influence over what brands do. On the contrary, the spectre of capital flight – the ever-present threat that brands will simply leave the country if over-regulated – is a constant looming presence in any such decision making.[6] Governments want the money provided by overseas production; brands do not want to be subject to interventionist government regulation. So, unlike the first scenario, in a globalised production process, environmental management becomes a game of whack-a-mole: environmental impacts dealt with on an individual basis, and only when they become especially egregious, rather than leading to structural change.

At this point, the sceptical reader may be tempted to raise a query along the following lines: 'OK, so governments in producer countries don't have much capacity to regulate and factories aren't greatly incentivised to self-regulate, but why don't the brands simply *inspect* their factories, to avoid these kinds of practices?' This is a reasonable point and the short answer is that some brands do indeed do this, for some factories, in some cases. To communicate why such measures may not be as effective as one might think, however, we need to eschew the neat

Carbon colonialism

'brand's-eye-view' of a supply chain, for a 'human's-eye-view' that brings us closer to the realities on the ground.

This is not as easy as it seems because we have all, in recent years, become used to the idea of globalisation. The apparent compression of space brought about by innovations in transport and communication has been called the 'Death of Geography': the idea that distance no longer has the same meaning in the face of technology's galloping pace. Holidays that span the globe have become the norm for many: the 30,000 km round trip from London to Bangkok no longer an exceptional journey, but for many an annual vacation. The miracle of a video call between Singapore and New York, similarly – a conversation that criss-crosses a hemisphere in real time – is for millions of people a regular part of the working day. We've become so used to these everyday challenges to the long-standing laws of distance that we have internalised their demise.

Yet, as various scholars have pointed out,[7] this death – like many others proclaimed amidst the white heat of Modern progress – has been greatly exaggerated. For many people, it has become normal to see the world in terms of flight times: five hours from London to New York, two hours from Paris to Berlin, twenty hours from San Francisco to Sydney: unimaginable distances traversed in the time it takes to watch a few movies. And major capitals like these may indeed have been connected, linked by high-density networks of flight paths and fibre optic cable, but beyond these gleaming intercontinental bridges, distance remains very much alive.

What is different now is that maximum speeds across distances have diverged substantially from the ordinary capacity to traverse them. It may well be possible to travel from an isolated factory eight hours from the Cambodian capital Phnom

The myth of a sustainable future

Penh to a brand's head office in London, but the number of people who have ever made the trip, in either direction, is minimal. What is forgotten in the conceptual realm of distance is the economic and logistical practicality of journeys that may be theoretically possible to achieve, but impractical, expensive, and – crucially – of little value in terms of information exchange.

Take the example of Tu, clothing supplier to the UK supermarket Sainsbury's. Tu sources from 312 factories in 13 countries. To continue with our earlier example, 24 of these factories are in Cambodia.[8] Yet Tu, like many major brands, does not have a physical presence in Cambodia. They have no office, no staff, they own no factories. Instead, they subcontract orders from partner factories over which they have no direct control. If a brand like Tu wished to check up on the consistency of their supplier factories with their parent company's environmental commitments to 'cutting carbon in our operations and using new technologies to maximise energy efficiency',[9] then they could, in the first instance, ask the factory what they were doing. This is extremely simple; it would involve only picking up the phone. If they actually wanted to *check* what was happening, though, consider the logistics involved.

After a fourteen-hour flight from London to Bangkok, followed by a further one-hour flight to Phnom Penh international airport, the journey begins. Our hypothetical Tu inspector drives their hire car towards the edges of Phnom Penh, where urban development begins to thin. Noisy density begins to give way to new low-rise housing developments, still under construction. In between is scrubland, trees, the occasional roadside stall, often selling only petrol. The drive to the most remote factory from the capital, in the Northwestern province of Battambang,

Carbon colonialism

is cited in travel guides as five hours, but perennial roadworks mean it usually takes at least eight hours of slow, bumpy, dusty travel, after which the inspector will likely check into a hotel in the provincial capital before beginning the hour's drive to the supplier factory the next day.

Finally arriving, they will pull up outside a large roadside compound surrounded by high walls. After being greeted at the front gate they will be led into an air-conditioned management office, furnished with high-backed leather chairs and a large wood veneer table. The management, like that of the majority of garment factories in Cambodia, is foreign, most commonly from China (in over half of cases), Korea, Hong Kong, or a range of other wealthy countries in Asia and the West. Generally, they don't speak English themselves, so the Tu inspector will be told of the company's environmental credentials via a friendly translator. A tour of the facility will be arranged, involving a walk around the sewing floor, the warehouse, perhaps a cursory examination of waste disposal. Hundreds of workers will be attending to their labours under the fluorescent lights. The facility will look clean, but the tour will be noisy with machinery, it will be very hot, it will be difficult to grasp much beyond the explanations being offered. The aspects of production under investigation will usually be in order. A smile, a handshake, a departure, and on to the next of the twenty-four factories. Including travel time, this one took three days.

This is a fictional account, based on many years of experience of similar visits, as well as conversations with those who do them regularly. It did not take place, because for most brands it rarely does. Indeed, this is what the story is intended to indicate. In a global supply chain, especially one in which high

The myth of a sustainable future

levels of competition mean that costs are paramount, the difference between making arrangements and checking on those arrangements is so vast that inspections are often dispensed with altogether. Distance and the practical logistics of geography still matter: they shape what is known and how well it is known, they shape the questions that are asked. Above all, they shape the difference between observed and assumed reality. The greater the distance, the more difficult the journey, the weaker the true connection between those in charge of a supply chain and those in charge of the everyday business of sustainability.

Ignorance is green capital

All of this points to a key underlying truth about the environment in a globalised world: that we know far less about it than we think we do. Yet what we have touched on rather less so far is a second core truth: that this ignorance is extremely lucrative. From misleading national carbon footprint reduction figures to false claims of supply chain sustainability, the ability to pass off environmentally destructive economic processes as clean, or at least as becoming cleaner, is a resource of high and growing value in a world in which environmental breakdown is increasingly difficult to ignore. This ability to sweep the uncomfortable, the unpalatable, the dirty, and the dangerous under the murky carpet of global production was always a feature of colonialism. For colonising countries like Britain, 'nature was a blank slate, to be reconfigured and rendered useful',[10] so what happened in the colonies, by and large, stayed there. Our present, unequally degraded world, is not simply the legacy of this system, but its heir in more ways than one. First, the global

Carbon colonialism

factory continues to push waste to its margins in ever-greater volumes, but more important still is the second dimension of carbon colonialism: control over knowledge of our global factory. That it is still, in other words, possible to proclaim the system just, clean, and fair.

It is this inherited capacity to conceal and misdirect that is perhaps the greatest boon to opponents of meaningful action on climate. Gone now, after all, is the era of the 1970s and 1980s, when the scientific evidence for climate change was meaningfully debated by scientists themselves. Gone too is the era which surpassed it in the 1990s and early 2000s, when doubts were sown amongst politicians and the media, predominantly by interests linked to the fossil fuel lobby,[11] over whether we could be *sure* that humans were the cause of this warming, or whether it might instead be rooted in the kind of changes which had shifted the climate in previous centuries: the relationship between the Earth and the Sun, or 'naturally' produced greenhouse gases through volcanic activity.[12] These red herrings served their purpose of delay and distraction for a period, but as arguments like these were systematically debunked, the twenty-first century saw the dawn of a new era of climate consensus: one in which the scientific agreement was bolstered by high-profile empirical evidence. The lethal European heatwave of 2003, Hurricane Katrina in 2005, Cyclone Nargis in 2008: none of these events confirmed the idea of climate change in itself, but as their regularity increased, alongside temperature records toppling with morbid inevitability each year, the sense of climate change as a present, tangible problem shifted the terrain of environmental debate. That climate change is here, that it is dangerous, and that it is getting worse could no longer be denied.

The myth of a sustainable future

Amidst this new landscape of public opinion, the vast collage of actors and interests comprising the global economy have had little choice but to effect a shift in approach. Public pressure to *do something* has become irresistible, so politicians and corporate actors must show that they are indeed doing something. Yet politicians and corporate actors are not beholden to the public alone. Shareholders, investors, even the public themselves demand economic expansion as well as environmental sustainability. This is far from easy. Various theories of how such expansion may be achieved at the same time as a reduction in carbon emissions have been posited, some – like Nobel Prize-winner William Nordhaus's green growth modelling – to great acclaim. These will be discussed in more detail later in the book, but the reality is that empirical evidence of their effectiveness at a global level remains scant. The inconvenient truth of the day is that if we look at the planet as a whole, economic growth and environmental degradation continue to go hand in hand.[13]

On the level of corporate governance, the solution to this is more or less what we have seen above. From a business, or indeed a political perspective, it is not necessary to be sustainable, but rather only to *appear* sustainable, a phenomenon amply demonstrated in decades of corporate 'greenwashing'.[14] Yet whilst early efforts at greenwashing were often simply untrue – such as Chevron's infamous F-310 petrol that fraudulently promised to turn 'dirty emissions into good clean milage' – our new globalised economy presents far more sophisticated opportunities to appear sustainable with minimal effort. A 'don't ask, don't tell' approach to long and complex supply chains is, as outlined above, quite sufficient to hide many industrial impacts on the environment, making the appearance of sustainability easier than ever to achieve.

17

Carbon colonialism

There is, of course, a fundamental flaw to this approach. Without meaningful global action on climate change, its impacts will continue to be seen and felt ever more strongly in the years ahead. Yet, as outlined above, this is not the problem it might seem. National wealth is as effective an antidote to the threat posed by climate change as any available, presenting a perverse incentive towards continuing economic expansion, even at the cost of sustainability. In effect, you can either stop contributing to environmental destruction, or you can accumulate the resources to mitigate its impacts. The evidence thus far suggests that you can't do both.[15] And the evidence of increasing resource extraction and accelerating global emissions suggests that it is the latter path that continues to be chosen.

In this context, healthy and secure environments are becoming increasingly scarce and unequal resources. The environment of many rich nations continues to improve, whilst the environment elsewhere degrades on an opposite trajectory. These are not independent processes, but a reflection of a global economy in which the dirtiest and most destructive, but still necessary, industrial processes that underpin our way of life have been exported to countries which must also bear the longer-term brunt of their global impacts. Contemporary global supply chains increase global environmental risk through carbon emissions whilst also siphoning the resources necessary to deal with these risks. Rather than environmental progress, this is environmental trade.

The remaining chapters of this book will outline the systems and processes that support this arrangement in two sections. The first of these, 'Greenwashing the global factory', will outline the environmental fallacies of globalised production, puncturing the comforting but misleading narratives of environmental

The myth of a sustainable future

progress that shape so many aspects of globalised production. Aiming to reset how you view the relationship between economy and environment, it begins first of all with the crucial question of how we got here, outlining the history of globalisation through the lens of the environment in chapter 2. As it will explain, changes in the global landscape of production in the last half-century have shaped the experience of environmental change around the world. Resource extraction and environmental degradation have followed broader processes of economic development, but they have also shaped them. Degraded landscapes have instigated social and economic changes, in some cases deepening the inequalities that facilitate environmental degradation and in others mobilising extraction into new territories.

Having set up the wider context that underpins the present situation, chapter 3 will consider the ever-present question on consumer lips the world over: what can we *do* about this? In this chapter, I will delve deep into the question of consumer power and consumer knowledge, exploring how the length and complexity of supply chains create a cloak of invisibility around production. Touching on the histories of greenwashing and the global factory, this chapter will shine a light on the knowledge gaps undermining claims of sustainability in our complex world of international logistics and container freight. Ultimately, it will question how much knowledge we as consumers really have about the mechanics of production we depend on and the steps we need to take to push back against greenwashing in the global factory.

Having explored how global supply chains hide their local environmental impacts, chapter 4 then zooms out to show how global production conceals carbon emissions on a grander scale.

Carbon colonialism

Carbon emissions are the greatest environmental threat facing the planet and have – at least on paper – been subject to ever more stringent regulation in recent decades. Major economies have made significant strides in changing the direction of their emissions, beginning to bend down curves that had strained ever upwards for centuries. The EU's net emissions fell from 5.6 billion tonnes of CO2 in 1990 to 4.2 billion in 2018, whilst the UK – historically one of the EU's largest emitters – claims a 44 per cent reduction in emissions since 1990: a world-leading achievement. Yet all is not what it seems. Elsewhere in global supply chains, emissions on a grand scale are hiding in plain sight, shunted to the global margins and away from carbon accounting mechanisms. This, this chapter will argue, is carbon colonialism.

Having outlined the ways in which carbon emissions and environmental degradation are hidden in the global economy, the second part of this book will explore how our globalised economic system structures the impacts of climate change. As this latter half will show, vulnerability to climate change is not an inherent characteristic of people and regions, but the result of unequal economic and political power. To exemplify this, chapter 5 turns to the issue of labour, a vastly underappreciated aspect of industrial sustainability. As it shows, vulnerability to climate changes is not only about changing environmental conditions, but also the resources people have to meet those conditions. The simple fact of having less money means greater vulnerability to all aspects of climate change, from storms and floods to food shortages due to crop failure. With many export-oriented overseas workers increasingly vulnerable to climate change, global supply chains are therefore playing a dual role in driving climate change impacts: on the one hand contributing

20

The myth of a sustainable future

to the ratcheting pressures of rising temperatures, unpredictable rainfall, and rising tides, but also placing workers at the front line of their impacts.

Issues like these, this book argues, should be at the forefront of discussion over the sustainability of the global economy, but remain side-tracked by a persistent focus on national carbon accounting targets and sustainable consumption practices led by wealthy nations. In chapter 6, I discuss some of the underlying reasons for our persistent inability to change the narrative on climate action. As this chapter argues, the root of the problem lies in who gets to speak and who doesn't: the global inequality that underpins environmental thinking and undermines our capacity to rethink our relationship with the global economy. Whose words – and whose ideas – are valued and whose are not is not simply a question of who speaks the most sense. On the contrary, the right to speak, closely associated with wealth, is passed down through generations. In order to revitalise our politics of industry, we need to apply the same duty of care to the workers and environments that make our goods overseas as those that used to produce them domestically. First, though, we must reverse our colonial subjugation of their voice.

Yet, as chapter 7 shows, this is harder than it sounds. Taking a tour through the diverse alignment of people, groups, and opinions that make up the world of environmentalism, this chapter highlights the power dynamics and interests that shape the global climate conversation. Beginning at COP26 in Glasgow, it examines the fractured landscape of climate change and sustainability policy, from the financiers to the activists, in order to understand the interests that shape how we think about environmental breakdown. As it aims to demonstrate, the key battleground on climate change is no longer with those

Carbon colonialism

who deny the existence of climate change, its human origins, or importance, but the 'wolves in sheep's clothing' who advocate delay and diminution of vital action. In contrast to the loud condemnation of climate denial, this cold war bubbles beneath the surface of environmentalism, a fight for knowledge, policy, and discourse, but above all a fight for the middle ground: to control what constitutes prudent climate action in the public imagination. It is against these false voices of prudence that the greatest battle remains to be fought.

Building on the preceding seven chapters, the final chapter will conclude with the book's overarching point: that environmental security – meaning physical safety from the impacts of natural and human-made environmental hazards – is a scarce and wasting asset that is becoming more unequal in its global distribution, diminishing rapidly in the global South and much more slowly in the global North. Under this system, the funds needed to cope with diminishing environmental security are obtainable only through means that worsen those impacts. In effect, climate change impacts, including the slow-burn disasters of droughts and floods, are being traded out by wealthier countries and imported by less wealthy ones as the price of economic growth. It is this vicious cycle that underpins the global economy under climate change, yet our inability to see this is underpinned by six key myths – from the power of sustainable consumption to the geography of climate vulnerability – that undermine genuine action on the climate crisis. Outlining each in turn, this final chapter offers the tools to think differently in order to make real change in the world.

Part I

Greenwashing the global factory

2

Founding the global factory: the first five hundred years

Around the spring of 2015, I found myself a passenger on the newly reinstated train between the Cambodian coastal towns of Sihanoukville and Kampot. Through the window rolled the spectacular landscape of the Southwest coast: densely forested hills, mangroves, and the distinctively spherical foliage of Cambodian palm trees along the track. It is an idyllic scene, rendering the looming grey concrete factory in the foreground all the more incongruous. More so still the denuded mountain it abutted, now raggedly half missing, crumbled into dust by remorseless machinery to fuel the breakneck concrete growth of the nation's cities. Listening to the distant clanking of steel on stone over the clattering of the train track I was reminded of the many Cambodian proverbs involving mountains. They mostly follow a similar theme of humility and knowing one's place: 'You, with the small arms, don't try to embrace the mountain', or 'don't throw your fishing line across the mountain'. The mountain always stands for power, authority, and permanence, making this one's sudden dismantling seem poignant, even undignified.

Scenes like this tend to bring out a predictable set of remarks from a Western audience and in this case it was no different.

Greenwashing the global factory

My fellow travellers pronounced the usual objections: 'how could people do this? How can they have so little respect for the beauty and value of this natural environment?' The responses of Cambodians, on the other hand, are mixed. Some will decry it as a travesty, others are more sanguine, lamenting the destruction but acknowledging the need for development and industry in the country. Whether interpreted as an aberration or a sacrifice, though, the crushing of a mountain is viewed inevitably as a modern phenomenon, made possible only by the harnessing of technology and fossil fuels which has come so completely and rapidly to reshape the natural world.

This is an understandable assumption. After all, the pace of environmental destruction has accelerated to alarming levels in the last few decades. Global material extraction, the basic removal of stuff from the ground, has more than tripled since 1970, from 22 billion tons to 70 billion tons.[1] And it is not only rocks, gems, and fossil fuels. Even as the industrial revolution began to change the world from the eighteenth century onwards, global forest loss remained relatively steady for decades, hovering at around 19 million hectares each year, before skyrocketing upwards during the twentieth century to a peak of 150 million in the 1980s. In the other direction, species decline has accelerated. Perhaps the rawest statistic of all is that, whilst the last half-century has seen the human population more than double, the global animal population has declined by 70 per cent over the same period.[2]

Nevertheless, whilst the world has seen environmental destruction on a different scale in recent times, the practices that underly it are anything but novel. It didn't, in fact, take the industrial revolution to endow humans with the power to destroy mountains, or create huge craterous mines in the earth.

Founding the global factory

The Romans were doing it two thousand years ago, somewhat more slowly, but with an equal disregard for the natural world. The Carrara Peaks in Italy and Mount Pentelekon in Greece still bear the scars of centuries of marble quarrying for successive emperors. On a more mundane level, the impact of Roman mines, which laid waste to local areas at the time, is still visible today. The fuel needed for heavy industry such as the iron centre at Populonia in Tuscany, which dominated European iron production from the sixth century BC to the second century AD, is recorded as equivalent to a million acres of forest each year. And it was far from exceptional. There were many such centres around the empire, with pottery production creating its own wastelands to match. Empire, even then, was rooted in unsustainability.

This should perhaps not be surprising. After all, extraction and exploitation have always gone hand in hand, in part because a certain kind of workforce is necessary to do the extraction in the first place. Wherever it has been practised around the world, from the Brazilian Amazon to the Congolese cobalt mines, extraction has predominantly been undertaken by a specific kind of workforce, detached from the mainstay of society. Extractors are, in general, socially similar, of male gender, and segregated for extended periods from families and communities. Before they can begin to drive forward the least sustainable processes that continue to underpin the global economy, in other words, extractors must first be extracted from society themselves.[3]

In its most extreme incarnation, this relationship is very clear in the present day. Many of the world's least sustainable forms of work are undertaken by people either forcibly separated from their communities, forcibly retained in their workplaces,

Greenwashing the global factory

or both. Cobalt mining in Democratic Republic of Congo is a notorious example of this, with workers – usually male, often children – bonded into dangerous, toxic, and often lethal work for years at a time to provide minerals needed for high-technology innovations like smart phones and electric cars. Work like this comes at a vast carbon and human cost[4] and is made possible predominantly by labour so cheap, and crucially so exploitable, that it is effectively disposable.[5] Similar patterns can be seen the world over: from the brick kilns of Bangladesh[6] to the depletion of fish stocks by captive Thai fishermen,[7] abuses of people and their environments tend to go hand in hand.

What's more, this is not just a moral issue, of those most willing to harm humans being readier exploiters of nature too. It's an economic inevitability. Natural environments can't, in most cases, regenerate fast enough to keep pace with global demand for raw materials, so extractors have to constantly expand into new areas: cutting further, digging deeper, casting their nets more widely and indiscriminately. Yet there's an economic problem at the root of all this. We are used to the idea of economies of scale; of production getting cheaper the more you make, but when it comes to extraction the opposite is often true.[8] Whether digging deeper for minerals, or cutting new and longer pathways into the Amazon, growing scarcity leads to rising costs, encouraging the use of exploited labour as a way to keep profits in the black.[9]

On a global scale, the low profitability of extraction is one of the underlying reasons that former colonies have struggled to catch up with their former colonisers in the independence era. Most colonial economies were organised around extraction, providing the raw materials that drove imperial growth. As a result, even when the imperial administration is taken out, the

Founding the global factory

underlying economic structures put in place by colonisers are very difficult to get away from and continue to hold newly independent countries back. On a basic level, exporting raw materials adds less economic value to the country that does it than processing, manufacturing, and reselling those materials,[10] so for every watt of energy, every hectare of land, and every hour of work used to make goods exported from the global North to the South, the South has to generate, use, and work many more units to pay for it. For land the average ratio is 5:1, for energy it is 3:1, and for labour it is 13:1.[11] Extractive economies getting poorer relative to those that use their materials is, in other words, 'physically inevitable'.[12]

Physically inevitable. As Kampot's mutilated mountain slid out of view, the physicality of these global processes seemed unquestionable. Yet clearly this was not the whole story. Cambodia is not Democratic Republic of Congo. It has certainly seen its fair share of extraction, suffering one of the worst rates of deforestation in the world in the early part of the twenty-first century, not to mention two decades of such extreme overfishing and pollution on the ecologically critical Tonle Sap Lake that nets now capture 'more trash than fish'[13] according to monitors. Yet Cambodia – like many countries in our contemporary globalised economy – is, itself, a producer country, with the economic figures to match: averaging double-digit growth in a miraculous decade from 1999 to 2008 and making it the sixth fastest-growing country in the world in the first ten years of the twenty-first century.

Does this mean, therefore, that Cambodia, a developing nation with an increasingly dominant manufacturing sector, is now on the same trajectory of growth as its former colonisers: the far wealthier countries whose clothing it produces? For the

Greenwashing the global factory

last four decades, the consensus that it is has dominated international development. Yet, beyond the realm of theory, the number of countries successfully transitioning from low to high income is in reality extremely small: Greece, Singapore, and South Korea being the prime examples. In almost every other case, early promise has faded into what is known as 'the middle-income trap':[14] the difficulty, in technical parlance, of transitioning from a poor, extractive economy to a high-technology, mass-consumption, rich one.

Why this happens is the subject of much debate, but whichever angle you come at the problem from, what is conspicuously missing in all this is the flesh and dirt of experience. Industrialisation on a page may seem clean, scientific, even inevitable, but global economic integration in practice is a far uglier beast, involving vast and often painful upheavals in the lives and livelihoods of the people who live it. Some of these choices are made enthusiastically, but many, if not most, are deeply difficult, taken in constrained circumstances and often involving substantial personal sacrifice for the future benefit of others.

This is the lived experience of economic development, not only under climate change, but also more broadly under capitalism. The construction of a factory does not automatically bring a workforce, at least in the first instance. The transition to industrial work is, for the communities and people involved, a process of social, cultural, and emotional transformation. At the very least, it means sons and daughters leaving behind friends, family, parents, and children, often indefinitely. It means working under high and constant pressure, with very little respite on a daily, weekly, or even annual basis. It almost always means tolerating uncomfortable, even debilitating, conditions for

Founding the global factory

insufficient pay. For those involved, it is not a smooth, predictable transition; one more fraction of a point of GDP accrued on an annual account of national growth. For those involved, it is a rupture, a sacrifice, a limbo.

It is also a chance grabbed with both hands by many prospective workers, which begs the question: what are industrial workers escaping from, for industrial work, with all its difficulties, dangers, and discomforts, to be a refuge? Traditionally, this has tended to be one of the questions less asked in accounts of industrial development. The sense of the global South as 'terra incognita' persists through the changing historical terminology used to describe it.[15] The evolving labels of 'colony', 'third world nation', and 'emerging market' reflect profound geopolitical shifts, but at the root of all three is a sense of lands and people which the 'magic of the market'[16] might set free, transforming them simply on account of its arrival. Yet, despite centuries of persistence, the reality has never been this way. The history of industrialisation has been a story as much about the creation and maintenance of an industrial workforce as the tools and technologies that that workforce eventually agrees to take up.

The creation of an industrial workforce

Several days after my train ride through the embattled hills of Kampot, I was conducting rural fieldwork on the other side of the country: the flood- and drought-hit province of Prey Veng, near the Vietnamese border. Most of the conversations had proceeded as expected, on the subject of rainfall, or its absence, and the pressures of the banks to repay the latest round of microcredit loans. So, it was only after coming across several

Greenwashing the global factory

tarpaulins covered in neatly arranged rows of cow dung that a new question occurred to me. 'Are you selling these?' I asked the farmer I was speaking to. 'Yes, we sell it to the Vietnamese.' I found this interesting, not only because the existence of an international dung trade had up to then been beyond the scope of my imagination, but also because this and other farmers had spent much of the day explaining how the use of fertilisers made their rice crop inedible.

Faced with an increasingly unpredictable climate, underpinned by rainfall patterns that no longer followed the same traditional rhythm,[17] the farmers in this village, alongside many others like them in Cambodia and elsewhere, had turned to modern methods. Pesticides, chemical fertilisers, new quickgrowing seed types, mechanised harvesting: all these innovations were adopted by farmers to cope with the loss of their labour supply to the country's nascent garment factories, and a climate increasingly ill-suited to traditional agriculture. The result was resilient, it was modern, but it tasted awful; so bad that the farmers elected to sell the rice they produced with these modern methods as animal feed, to the same Vietnamese traders that bought the community's dung. With the money they received, they would buy rice they could actually eat.

So, farmers were exporting the means to make edible rice, whilst importing the means to make inedible rice. It appeared a curious arrangement, but what made it seem especially notable at the time was my own ignorance. This conversation took place in a village in the Southeastern Cambodian province of Prey Veng, in a district far from urban development and difficult to access for much of the year, due to waterlogging which turned the roads into mud baths. Until very recently, access to Phnom Penh was possible only by an ancient ferry imported

Founding the global factory

from India, which criss-crossed the Mekong throughout the day. I had once spent a night in this ferry port town, Neak Luong. The electricity went out at 9pm and didn't return until 5am, as in most of the province. To me, a foreigner from a wealthy part of Europe, it *felt* like the end of the world. And yet here, a further two hours' drive into a rural world of villages and emerald rice fields, the market – the international market no less – was an essential, inescapable, ever-present dimension of local ecology.

As I soon began to realise, this was not just a one-way process, not just modernity and development encroaching on the old ways of life. It was a cycle, in which the development of urban areas shapes rural changes, eventually making each dependent on the other. In Cambodia, this had begun – or more properly, begun again – relatively late in historical terms. Following the devastation of the Khmer Rouge period, when urban areas were largely evacuated and most industries halted as the population was conscripted into agrarian labour camps, Cambodia entered a decade of limbo under the Vietnamese-controlled People's Republic of Kampuchea.[18] Civil war with persistent Khmer Rouge groups continued to scar the country well into the 1990s, leaving Cambodia one of the world's poorest countries, with only a tiny industrial output to speak of. In the intervening years, the explosion of economic growth has been called a 'miracle',[19] yet it was to some extent predictable. A country geographically well located in global supply chains, in which a huge supply of cheap, mobile labour is suddenly made available to the global market, always had the potential to thrive. After all, a workforce of millions clamouring at the factory gates is precisely what investors want to see.

33

Greenwashing the global factory

But let's step back for a moment from the cold economic rationality of growth models to consider how this process actually looked to the people who lived it. The Cambodian garment industry in the 1990s was notorious. Child labour was common, sexual harassment – even sex for jobs – was rife, wages were extremely low, and conditions were often appalling. Yet, despite this, the workers kept coming. In the last quarter-century, Cambodia's garment industry, the engine and driving force of its economy, has grown in scale more than 200-fold.[20] It has expanded its workforce from a few thousand in 1994 to more than 750,000 today, from an estimated working-age population of around 9 million.[21]

Almost one in five working-age Cambodian women is currently a garment worker. And still, they keep coming, crowding patiently at the factory gates when a new batch of jobs, or even one job, is announced. Three decades into the industry's explosive development, this still strikes me as remarkable. Things are much better than the dark days of the 1990s, but this is still difficult, draining work. On a human level, what explains this sustained flood of people to the industry, this vast human wave of recruitment to an industry even now associated with long hours, malnutrition, and ill health? To underscore the importance of this question, let's think through the average day for a garment worker.

You wake at 5.30 in the morning, in a concrete windowless room, about ten metres square. There are between three and eight others sharing the room with you; as many as possible to keep the rent down. A single floor fan battles a heat that remains close to 30 degrees even at night for much of the year, higher in an airless space like this. You prepare breakfast, a thin rice porridge, for yourself and the others, before setting off to wait

Founding the global factory

at the roadside for the truck to take you to the factory. When it arrives, it already looks full to bursting, but you and another half dozen others squeeze on, joining fifty or more fellow workers shoulder to shoulder on the back of a flatbed truck. It is uncomfortable, but discomfort takes a back seat to fear as the truck careers through traffic on the treacherous national roads. With workers only held in place by one another's bodies, these trucks are extremely dangerous, with dozens of workers killed each year when they crash or overturn. Dozens more are badly injured, losing limbs to the asphalt. It comes as a relief to see the factory up ahead.

The working day begins at 7am, so you begin filing through the gates alongside hundreds of fellow workers. You reach your sewing machine and begin work. The pressure to work fast is intense. Your line manager shouts at you when you slow down even a little and any mistake will unleash a tirade of abuse. 'Work faster or get out' is a famous refrain.[22] Bathroom breaks are prohibited, so after four hours of constant labour you have sewn the right-hand sleeve onto hundreds of versions of the same shirt. At 11am, the bell rings for lunch, so you file out with hundreds of colleagues to the front of the factory where various food stalls are waiting. The food, rice with vegetables and a little meat, is cheap, but very poor quality. You still feel hungry when you file back with the other workers at 11.30am. By now it is extremely hot. The huge factory is simply four brick walls with a metal roof. There are no windows and the few air vents at the top of the room are inadequate. You feel nauseous and faint as the temperature rises. A handful of workers in the room do lose consciousness, but you struggle through a few hundred more shirt sleeves until 4pm, when the bell rings for the end of the day. You file outside again, grateful to feel the breeze for

Greenwashing the global factory

a moment, but today is a good day. An overseas order means overtime is available. Like almost all workers, you take it whenever it is offered, so you file back in for another two hours. Arriving home at half past seven after another death-defying truck ride, you begin preparing dinner, this time rice with eggs, before settling in for a deep, exhausted sleep.

This is how the day looks for millions of garment workers around the world, six days a week or more, year-round.[23] Long hours, ill health and a world constricted to a pinhole of experience; a life lived between the sewing machine and the walls of a rented room. Garment workers can cope with this kind of work for a period, but it is ultimately unsustainable. The vast majority expect to return home at some point and most ultimately do so. The average age in the Cambodian garment industry is twenty-seven. In Bangladesh, it is twenty-five.[24] This is by and large a young woman's game: something that people take on for a period before returning to rural life. Only the very poorest tend to stay long term and few are happy about it, complaining, as a garment worker once put it to me, of 'no future', and only 'work without end'. For anybody who can afford it, industrial work has become a phase of saving money, often prior to marriage, to support the family farm, or save money to invest in a rural business such as petty trade or cash cropping.

The problem is that this is not how industrialisation is supposed to go. Theories of development describe a process of permanent transition; the abandonment of subsistence farming for industry and industrial agriculture. Numerous models outline this process, like the economic doctrine of Convergence Theory, which argues that industrialising countries come to resemble one another, with poorer countries growing faster, so that everyone will ultimately end up in the same place.

Founding the global factory

Nowhere in these models is there anything about urban development and smallholder agriculture co-existing sustainably, of rural people remaining rural, supported symbiotically by industry. On the contrary, the opposite is true. The seminal Dual Sector model proposed by W. Arthur Lewis in 1954, which still underpins international development planning, was entitled 'Economic Development with Unlimited Supplies of Labor'. It is, in other words, hardwired into theories of development that a never-ending flow of people is on hand to fuel the appetite of industry.

You could certainly be forgiven for thinking this in Cambodia, as I discovered several years ago whilst working on a study of working conditions in the garment industry for an international NGO. During that study, I interviewed several factory managers to ascertain how long it took to replace a worker when they left the industry. My hope was that this would help bosses to see the economic value of retaining workers by improving the conditions they faced in the industry, but the results were disappointing. In fact, many managers did not even fully understand the question. 'It takes no time at all', they would eventually say. 'They are queuing up outside the gates. We just pick one from outside.'

So, let's recap. Garment work is hard and physically degrading, most workers wish to return to their rural lives, and many ultimately do so. Given this, what keeps people coming in the first place? Why do the crowds remain outside the factory gates? The obvious answer is money. In the 1990s, most Cambodians were extremely poor. Many lacked food security, access to medicine, access to education, even clean water in many cases. But we are no longer in the 1990s. Nutrition, schooling, and health have greatly improved. Rural and urban

Greenwashing the global factory

opportunities are far more abundant, so why do the crowds continue to gather? If people would prefer to stay, why don't they simply stay?

The answer lies in those neat lines of dung, destined for the international market. If they ever were, rural areas are no longer closed systems. They are integrated into the market and require capital to function. Take rice farming, for example. Traditionally, Cambodian rice farmers, like most of those planting the crop around the world, would practise a transplanting method of farming. This means painstakingly planting the seeds produced by last year's harvest in a small nursery plot of a few square metres and tending them carefully until they emerged as seedlings. They would then dig them up and replant them in a larger plot to grow to maturity, until they could be harvested. The seeds would then be carefully collected, and the next year the cycle would begin anew. Almost everybody in Cambodia farmed this way in the 1990s. Yet today, almost nobody does.

Rainfall has become sporadic and unpredictable and it is not worth spending a month lovingly nursing and replanting seedlings, only for an unseasonal period of drought or flooding to see them wither away. There are other ways: fast-growing seeds, fertiliser, pesticide, irrigation, mechanised harvesting, but these all cost money; a commodity in short supply in the planting season, with the proceeds from last year's harvest gone. So, farmers can either take out a loan or send a family member to work in industry, so that they can send something back for the farm. In reality, the answer is rarely one or the other. It is both. Debts can be taken out at any time, but they have to be repaid in regular instalments. Farmers soon find that lenders won't accept payment according to the rhythm of the seasons.

Founding the global factory

Monthly repayments require monthly salaries. And so both family and farm are absorbed into the market.

Once this process has begun, it becomes self-reinforcing. More people leaving rural areas means a falling supply of rural labour. The cost of hiring agricultural labourers goes up, making the old, labour-intensive way of farming financially unviable. So, more farmers start borrowing money for new seeds, fertilisers, and pesticides, sucking the nutrients from their plots year by year. Before long, chemical fertilisers are the only thing that can draw crops from their exhausted soil. Natural fertilisers – that dung we keep returning to – are a solution, but a useless one. They would replenish the soil, but nowhere near quickly enough to make those pressing monthly debt repayments. So, farmers instead line it up neatly on their tarpaulins, ready to be whisked away for a few dollars more chemicals.

And so runs the globalisation story for many around the world: environment and industry operating as symbiotic cogs turning the wheel of modernisation. It is in one sense a tale of its time, rooted in the thoroughly modern phenomenon of climate change. Yet this small account of the interrupted development of a troubled sub-tropical country of 17 million speaks to a wider and longer-term truth about industrialisation. 'Build it and they will come' has never filled a factory in a pre-industrial context. The change of life is too profound, too fundamental, and on a basic level too unattractive, to entice a workforce out of farming. Something else is necessary. Not just the financial pull of modern money, but a push to jump start the long and lonely journey from farmer to worker.[25]

There are many forms this final push can take, from economic shocks like the illness of a relative, to disasters like floods, droughts, or landslides. None of these pushes is new. After all,

Greenwashing the global factory

rural livelihoods have always been uncertain and risky. Yet in recent years, the pace, intensity, and power of the shoves and barges parting farmers from their land have grown. Droughts have gone from once a decade to most years, unseasonal rainfall now occurs every season, and debts tick steadily upwards. That last resort move away from a rural home, either for the whole family or more often just one member of it, has become no longer a means to earn and invest new income, but to sustain what was previously sustainable. The rural and the urban, the agricultural and the industrial, have become a hybrid system, dependent both on one another and the international market to function.

The textile industry, past and present

Driving back to Phnom Penh, over and around the craterous red dirt roads that would lead me eventually to the old port town of Neak Luong, I considered the implications of this rural transition. On the one hand, it seemed idiosyncratic, rooted as it is in a tragic and bloody history which saw Cambodia's industrialisation defibrillated from a far lower base than its close neighbours. Yet Cambodians now are the heirs to many of the same pressures borne by their forebears in the global textile trade, the 'dark satanic mills'[26] in which workers suffered unimaginable privations in service to the global 'empire of cotton':[27] an industry so central to the industrial revolution, and so profitable, that it fuelled Britain's rise to the apex of imperial power.

In his remarkable book on the textile industry's ruthless rise to world domination between the sixteenth and nineteenth centuries, Sven Beckert has outlined in depth how Britain and other industrialising countries forced the world, thorough

Founding the global factory

politics, power, and violence, to participate in a system of globalised trade with European powers at its centre. Even after a working life spent in large part concerned with the practices and abuses of the global garment industry, exploring this history surprised me. There is something almost inherently banal about clothing, an everyday quality that speaks to the idea of a trade that is intrinsically benign, but corrupted. Even the many toxicities of fast fashion have not – for me at least, and perhaps for many others – dispelled the notion that if only we could strip away the industry's more rapacious qualities we might be left with something sustainable, something *necessary*.

The problem is that, historically speaking, once you've stripped this all away you're left with nothing. British industrialists in the eighteenth century had developed ground-breaking new and exciting tools – the spinning jenny, Arkwright's water frame – this much is true. Yet what they possessed in technology they lacked in resources. Even if Britain had been a climatically suitable environment for cotton growing – which it was not – the area needed to cultivate the cotton it was spinning by 1830 was greater than the country's entire landmass. Only when the United States entered the market in the early nineteenth century, beginning an explosion of cotton production – from 1.5 million lbs in 1790 to over 2 billion lbs in 1860 – was the issue of supply resolved, but it was a solution built almost entirely on slavery. With one in thirteen Americans enslaved in cotton plantations by 1830, slavery, which had 'enabled industrial takeoff, was now integral to its industrial expansion'.[28]

All of this, though, was the easy part. Raw cotton still had to be woven and spun in Europe, where chattel slavery of the sort practised in America was unthinkable. This proved to be an immense industrial challenge for the simple reason that work

Greenwashing the global factory

in cotton production was difficult, dangerous, and appallingly paid. Workers were required to labour a minimum of twelve hours a day, six days per week, year-round in dusty, humid, and deafeningly loud conditions. Serious accidents, including the loss of limbs or fingers, were frequent, which combined with the overall physical degradation of the work, necessitated a constant supply of fresh labourers.

Child labour was, as ever in industrial Britain, a key part of the solution. In 1833, 36 per cent of Lancashire cotton factory workers were aged under sixteen, with many as young as eight.[29] Young workers like these had the advantage that they were cheap, being paid a quarter to a third of the wages of an adult man, and could be formally indentured on 'apprenticeship' contracts of up to seven years, during which they would toil, rather than learn, without the possibility of leaving to work or study elsewhere. Many never lived to see the end of their term. Of 780 apprentices recruited in one English mill in the two decades from 1780, 119 ran away, 65 died, and 96 had to return to their parents or overseers. Throughout industrialising Europe, the adult population fared little better. When the Saxon government sought to recruit soldiers from the cotton industry around the turn of the nineteenth century, only 16 per cent of weavers and 18 per cent of spinners were deemed healthy enough to serve.[30]

Statistics like these, or more accurately the tangible horrors they represented to potential workers, presented, as they do now, a major problem to recruiters. Not only was the cotton industry bad work, but it also represented a complete break from a traditional way of life. The only way in which people could be induced to participate in industrial labour was therefore to make traditional forms of work – far preferable by

Founding the global factory

default to the industrial alternatives – redundant, inaccessible, or illegal. This was done on the one hand by state legislation that criminalised non-compliance with the labour law, that is, running away in most cases. Yet it also famously revolved around the degradation of traditional rural ways of life through the enclosure of common rural areas and the solidification and extension of ownership over productive land. Vast new areas of land were incorporated into the productive, industrial economy, and large parts of the population with them.[31]

This is history. All of it happened a long time ago and very far from the Cambodian village I had recently departed, but if you were to sit down with one of the village's inhabitants and describe the story of English enclosure over a plate of sun-dried fish, he or she would likely find it quite familiar. He or she might even assume that you were referring to a neighbouring village, where similar processes have been ongoing for decades. Since 2000, large areas of land have been titled and previously common land allocated first to the state and in many cases subsequently to private owners.[32] Market forces, just as they did in eighteenth-century England, have begun to disrupt smallholder farming. In just over twenty years, landlessness has risen from near zero, to 29 per cent of agricultural households,[33] many of them now queuing outside a garment factory's gates, a ready replacement for a worker either unwilling or unable to continue.

Industrial workers, in other words, are not simply found, but made. To obtain and maintain an industrial workforce requires, in its early stages at least, a pressure on traditional livelihoods to overcome the natural reluctance to enter industry. Once this is understood, the crucial interconnections between industrial and rural development become clear.

43

Greenwashing the global factory

Many strategies, from irrigation to seed banks to schooling outreach, can and are employed to make life better for rural people in the short term. Yet, taking a broader view, these strategies are by no means intended to make rural life sustainable for smallholder farmers. On the contrary, doing so in a true sense would be to work at cross-purposes to the needs of industry. Without workers flowing in from rural areas, the sewing machines would fall silent.

Climate change is therefore not only the end point of the industrialisation story, but also a catalyst for the expansion of the global factory. Each year, the deepening pressures on rural livelihoods swell the crowds outside the factory gates a little more. Each flood, each drought, each unpredictable period of rainfall increases the pressure still further on urban workers who have to support their rural households. What was once sufficient for personal and family needs may look meagre in the harsher and less predictable environments brought about by climate change, meaning reduced quality of livelihoods, longer working hours, and a greater vulnerability to exploitation by employers. For millions of people around the world, therefore, the changing climate does not mean simply changing weather, but worsening terms of work.

As my car approached Phnom Penh, this seemed abundantly clear. The continuum between the rural and the urban played out at the side of the road as if animated in real time. Concrete roadside shops slowly replaced wooden ones, rice fields melted into new residential developments. It is a transition that brings home the artificial nature of the binaries that shape our thinking. Rural and urban, farming and industry: to those who live amongst these changes, these processes are not only linked, but integral to one another. The old world doesn't disappear in the

Founding the global factory

face of the new, it isn't cast aside or deleted. Seen in real time, the two worlds instead enter into an ongoing dialogue, a story of compromise and innovation. This means a garment worker's wages to buy fertiliser for the farm; a sack of rice for a construction worker so they can send back more of their monthly wages to their rural household; a busy retirement for grandparents caring for infants whilst their parents work in the city. This is the lived experience of globalisation.

Ultimately, though, this is only an intermediary phase, a resistance. What is underway beneath the surface, indeed what industry requires to function, is a gradual decoupling of these integrated wholes, the recrafting of livelihoods and environments in a way that better fits the schematic of growth in the global economy. This is as true in twenty-first-century Cambodia as it was in eighteenth-century Britain. Then, as now, it means people separated from land, it means dung removed from the fields that nourished it, rice removed from the bodies that planted it, and trees cloven at the root.

This process of dissection, separation, and extraction has been labelled 'the death of nature' by Carolyn Merchant,[34] a story that she argues began with the scientific revolution of the seventeenth century. What started as an analytical disentangling of holistic ecosystems soon became a physical imperative, with nature subverted to the needs of the economy in a way lethal to both in the long run. Biodiversity is inherently messy, complex, and holistic. It is interdependent in a way that defies disaggregation. Yet only when this complex mess is unpacked can it be allocated a proper economic value. Like the dying mountain with which we began this chapter, rural environments and the livelihoods that populate them must first be destroyed before they can be reintroduced to the market as

45

useful commodities. The matter remains, but in new and malleable form: the stuff not of proverbs, but of economics.

The environmental history of industry is thus one not only of building up, but also of breaking down: of people from nature, nature from itself, and of value from culture. And this, in essence, has been the story of the global factory in the last five hundred years. Setting up an economic system in which everyone and everything is now embroiled did not come easy. People did not simply leave traditional livelihoods by choice in many cases, nor did they voluntarily give up their land for economic uses. From an early phase of expansion backed by force, globalisation has now arrived at a self-sustaining, or indeed, a self-degrading phase. In this present era, economies at all scales have become so mutually entangled that sustainability, in a local sense, is almost impossible. As, indeed, it must be for the system to persist.

Without materials being dug up, cut down, or moved from one place to another the wheels of growth stop turning entirely.[35] Whole global infrastructures, whole societies, are structured around the imperative to find fuel for the global engine.[36] It is a logic that has permeated lives and livelihoods to such an extent that extracting is barely even a choice anymore. Chemical fertilisers are not a poor choice but the only choice once they have rendered dung obsolete; dispossessed or over-indebted farmers travel by night to log protected forests because they have no other option. Fishing with electric poles is the only path left when trawler nets have swept a lake clean. Sustainability, in a word, is fragile, but it is also fundamentally local: anathema to the extractive logic of a globalised world. We are living in a global society founded on and rooted in extraction. Environmental degradation is not a by-product of

Founding the global factory

this system, but the engine of a machine that separates and sucks in materials, before exporting and returning waste.

Viewed in this light, the connection between environmental vulnerability in one place and safety in another becomes clearer. It is resources, above all, that are needed to tackle climate change. Those resources have been flowing in the direction of the same rich nations for centuries and are doing so faster each year, tripling over the last four decades.[37] In the other direction, the global waste trade has expanded exponentially since the turn of the millennium, from 98 million USD in 1988 to over 2 billion USD today.[38] These flows, in both directions, are extractive processes with colonial roots, the product, as Benedetta Cotta phrases it, of an 'imperialist mindset' in which the only way for global Southern countries to develop is through ever-deeper integration into economic arrangements that exacerbate environmental vulnerabilities for the poorer and less powerful parties.[39] Far from alleviating the impacts of climate change through development, therefore, this is a systematic outsourcing of climate breakdown: a redrawing of global risk to suit the interests of rich nations. In short, it is carbon colonialism.

3

Consumer power in the global factory: a lucrative illusion

One morning in the autumn of 2018, I was showing a group of attendees around an exhibition in London's Building Centre. Housed in the basement of a site used primarily for trade expos, the show brought together photography, worker testimonies, and data from a project called Blood Bricks, which had explored how climate change increased the risk of modern slavery in the Cambodian brick sector. Upon release, the report had garnered substantial media attention, reported in the international press in a way that piqued the public interest. For a postdoc working on his first full-scale academic research project it was disorientating, a juddering contrast between the mud and dust of months of fieldwork and the sudden interest of a well-heeled Western audience, most of whom had never been anywhere near Cambodia.

Less than six months previously, I had been sitting amongst brick workers, sharing the noxious smoke of the kilns, dusting off clothing labels trodden into the ground by thousands of hurried, weary footsteps. I had spent weeks at a time criss-crossing rural villages back and forth to identify survey participants, sweating until my shirt was soaked through as the hot season

Consumer power in the global factory

reached its apex. I'd watched in awestruck admiration at workers relating biographies of transcendental horror with impossible composure. It was a period of fieldwork hyper-real in its intensity, a million miles away from the staid grey bustle of an autumnal central London. And yet, here it was, blown up in high-res technicolour for the British public to digest.

I had at this point been working as a researcher in various capacities for a decade: first as a graduate researcher, then as a PhD student, a consultant, and finally as an academic. I had conducted dozens of projects, but the interest had always been local, sectoral, specific. So when, upon the conclusion of the tour, a kindly older gentleman and his wife looked at me and in earnest entreaty asked: 'so what can we *do* about this?' I was taken wholly by surprise. After a pause, I replied with the only answer I could give. 'Nothing', I replied, to the obvious disappointment of the audience, the confusion of the older gentleman, and the amused embarrassment of the colleague who had been leading the tour with me. Looking down at the floor in annoyance at myself, I resolved to attempt a better answer. Yet hard though my eyes scanned the carpet weave, I could not envision the solution that my questioner was expecting.

Leaving the exhibition later, I reflected on a place I had visited a few months previously: a clutch of unusually old brick kilns towards the Western outskirts of Phnom Penh. It was a part of town with a disturbing recent history. During the 1990s, it had been home to a notorious brothel called K11, so named for its 11 km distance from the centre of the capital. At a time when Phnom Penh was famed for its lawlessness, the Svay Pak area in which K11 traded had a national reputation, still recalled in bawdy drinking songs to this day. It had taken until the early 2000s to begin to shut down K11 and

Greenwashing the global factory

the reputation of the area has only recently begun to dim in the public memory of a young country. Yet underneath all of that had lain another open secret, which continued unabated to the present: Phnom Penh's most central clutch of brick kilns, a smoking sore of rag-strewn toil, well within the boundaries of the nation's capital.

What made these kilns so remarkable was not the scale of the operation. A group of four kilns is by no means uncommon, dwarfed by the hub of dozens of large-scale brick factories to the North of the city. Despite this, it stood out for two reasons. In the first instance, these were the old, round-backed elephant kilns that are now rarely seen around Cambodia. Standing around forty feet tall, they are fed with fuel via an igloo-like curved opening at the rear. The flat front is split by a door at ground level, topped by a long two-foot slit reaching the upper summit of the kiln. It is through here that workers enter to stack fresh bricks for firing first at ground level, then climbing higher and higher to continue building the rick. Once bricks begin to be stacked above head height, workers climb up the kiln, throwing bricks from the ground to a worker standing half-way up the slit and then in, to those standing atop the pile of bricks being built. It is dangerous work, not only for the ever-present risk of falling, but for the particular moment when the plates plugging the front of the kiln are removed by hand. If the burn has been mistimed, the blast of super-heated air and flame will exit the narrow slit at the kiln's front at high velocity, killing them instantly. Some of the workers had seen it; all had heard of it and were aware of the potential danger they faced.

Yet unusual as they were, it was not the kilns themselves that stood out, but the thick black smoke they were producing.

Consumer power in the global factory

Although many kilns burned garments as fuel, very few kilns burned only garments, and yet here, in an area now well within the boundaries of the city, were four kilns doing just that. Piles of rags, clothes, plastic bags, and labels were stacked high around the rear of the curved brick monoliths, oozing black fingers of molten plastic slowly working their way to the ground as plumes of pitch-coloured smoke swirled at eye level around the kiln, across the road, into the homes of the neighbours living just a few feet away. After my first trip, I had returned with a PM2.5 air quality monitor in an attempt to gather data on these fumes, but it was little use. Even standing across the road by the houses flanking the kiln, it shot to the maximum reading, 999, and remained there. Quite literally off the scale.

For the neighbours, who lived amidst rolling plumes of smoke for much of the day, this was naturally the cause of much concern. Burning garments is illegal in Cambodia. They had complained to the police, they had complained to the local authorities, but to no avail. The owner of the kilns was a well-connected person with the means to make problems go away, and so it persisted, unresolved and undiscussed. How, then, might the gentleman I had spoken to at the exhibition succeed in doing something to draw public attention to the issue? He might speak to his neighbours, show them pictures of the embattled Cambodian brick workers he had seen that morning. He might write to his MP, who might raise the issue in Parliament. Perhaps it would spark a debate on the travesty of this overseas injustice. A statement might be made by the Foreign Minister, or even the Prime Minister, condemning the conditions that Cambodian brick workers were forced to endure. It might even be taken up by the UN High Commissioner on Human Rights, who might make it a priority case for action.

Greenwashing the global factory

All of this is possible, but very unlikely. Not least because the world contains far more than a single example of cruelty and injustice. For the suffering of any given group to achieve centre stage, it must outcompete many similarly deserving others. And in any case, it was not a lack of attention, not the absence of will, that was blocking my mental pathway to a solution. On the contrary, all of this was far from a secret – barely even an open secret – and yet persisted nonetheless because, as in many such situations, problems are resolved by reducing their visibility rather than tackling them at the root. In the wake of the *Blood Bricks* report, for example, the initial government response had been to deny the report's findings via the national press. The claims of child labour, in particular, had stung, resulting in denials and refusal to engage. A year or so later, it emerged that certain actions had been taken: banners erected in a handful of kilns stating 'no child labour' and fences between accommodation and machinery. Kiln owners were also told not to speak to or engage with journalists, NGO staff, or researchers, or to allow them to enter their sites. Lip service had been paid and the issue had been shut down.

Whilst this might seem like a criticism of Cambodia, or its response to the issues uncovered, it is not. As one of the UN's forty-six Least Developed Countries, defined as having 'severe structural impediments to economic growth',[1] the Cambodian government has limited capacity to monitor industrial activity in detail, or even to enforce environmental statutes. The law acts as intended in some cases, as a deterrent in other cases, and not at all in others still. Yet this is not wholly dissimilar to the global North, where citizens are also wearily resigned to big business bending the rules. In large part, it is simply a case of big business looking smaller to a foreign eye than it does to

Consumer power in the global factory

those in the local context. The brick industry is both highly lucrative and supportive of wider national development aims. Construction in Cambodia has skyrocketed in recent years, with over 48,000 projects approved by the Ministry of Land Management in the last decade and some 53 billion US dollars invested.[2] Without bricks, there are no buildings, foreign direct investment falls, and the upward urban surge so closely associated with successful development around the world would grind to a halt. Would any country so reliant on an industry in which labour exploitation is rife choose to shoot themselves in the foot at this fragile state of development?

The historical examples are few and far between, but the ideal of moral boundaries delineating the acceptable territory of trade remains highly influential to this day, underpinning, amongst other examples, the development of the UK's Modern Slavery Act in 2015, which directs that companies publish declarations on the absence of modern slavery from their domestic operation and wider supply chains.[3] Despite the fanfare with which this Act was received though, its direct implications – as with other forms of supply chain regulation – are limited by the length, complexity, and obscurity of many supply chains. It is all well and good declaring that a supply chain must be free of something, but if you get to choose which bits of the supply chain count as 'yours' and to view them at something of a distance even then, that declaration may not mean as much as it appears.

Let's return for a moment to brick production in Cambodia as an example. There are around ten thousand workers living and working in the Cambodian brick industry. The majority of them are debt bonded, banned by the brick kiln owner from leaving or taking work elsewhere. The population of the brick industry includes almost four thousand children, not all of whom work, but many of whom do, either all the time, or on occasion, when their parents most need them to.[4] Not long after the publication of the *Blood Bricks* report, a twelve-year-old girl lost her arm in the kiln, but she was by no means the first such case. Every kiln has similar stories and only a few months before I had met another girl who had lost her arm in a similar way. I've met the mothers and fathers, husbands and wives, brothers and sisters of brick workers who simply died in their sleep in their thirties, forties, fifties, unable to endure any more.

You might wonder what these distant injustices have to do with Britain and the Modern Slavery Act. After all, this is an industry with, on the face of it, few international implications given that the bricks it produces are used entirely domestically. Yet in a world interconnected by flows of capital and foreign investment, the picture is far less certain. British companies invest in these buildings, they part own them, and they use them. At the time of the report, Black Rock, Standard Life, and the Norwegian Sovereign Investment Fund all had holdings in buildings constructed with materials produced by child and bonded labourers.[5] Yet indirect connections such as these are not covered by the Act, which holds that 'there is currently no specific requirement that firms need to consider their financial investments in their modern slavery statements'.[6] Huge profits are still attainable from the very practices specified by the Act; in other words, as long as they are obtained indirectly.

Nevertheless, financial flows like these are perhaps something of an oblique example of such connections, so let's consider a more direct one. To many readers, it may come as a surprise that, unlike in the Cambodian case, bricks are not an inherently domestic commodity. Despite their low value and

Consumer power in the global factory

heavy weight, billions of these baked clay lumps are transported around the world each year, either to plug a gap in supply, or simply because they can be made more cheaply elsewhere. The world's biggest importers of bricks are a varied bunch, comprising, in top five ascending order: South Korea, Saudi Arabia, Japan, Rwanda, and the UK at number one.[7] The UK imports, depending on estimates, between 14[8] and 16 per cent[9] of its total brick stock. The majority of these bricks come from Europe, with kilns in Belgium and the Netherlands producing most of the over 400 million bricks arriving on British shores each year.[10] Within this figure, though, a small but rapidly growing minority come from much further afield, with 33 million arriving from outside the EU, including almost 10 million from India and 14 million from Pakistan in 2019. The proportion of bricks imported from outside the EU increased more than tenfold between 2015 and 2019, from 3,088,902 to 32,942,280: a 59 per cent annual growth rate that, if continued, would see non-EU bricks surpass EU-produced bricks in under five years.[11]

That's some trend, but we need to dig into the data a little deeper to get to the nub of what it means. Non-EU, after all, is a big place. So where do these bricks actually come from? As it turns out, the majority of non-EU bricks, over 30 million in 2019, come from two countries: India and Pakistan, which collectively form the core of what is known as the South Asian 'brick belt'. In terms of both the ethics and sustainability of these brick imports, this is not good news. Brick production in this vast area of 800,000 kilns stretching from Afghanistan through India, Pakistan, Bangladesh, and Nepal is notorious for its hardship. The labouring population working in brick kilns often consists of some of the poorest and most marginalised members of

55

Greenwashing the global factory

the population. Child labour is widely prevalent, workplace hazards are common, and living conditions are generally poor.[12] Indian brick workers, who now make over 20 million bricks a year for the British market, have been reported by the BBC itself as 'living like slaves',[13] subject to abuse, exploitation, and physical and sexual violence inflicted by owners and managers upon workers.

Not long ago, I led a project investigating this growing trade in South Asia, in which we spoke to dozens of workers in Bangladesh and India about their experiences of work in these kilns. One worker described his experience working in the brick industry as follows: 'I feel weak due to the heat from the fire. My head gets hot. My skin has deteriorated as well. I feel terrible breathing in the fumes from the burning coal gas. They can find coal debris in my body when I get check-ups done, so I get coughing fits and colds as well.' This is, in other words, a debilitating, unhealthy, and unjust job, in which debt-bonded workers produce cheap bricks for citizens of some of the world's wealthiest countries. Very few consumers would be willing to buy bricks that they knew were made in conditions like this, but identifying the source of bricks is far from easy. The brick supply chain runs through a long and complex chain of middlemen and intermediaries, with little to no public oversight. Tracing the path of a given brick from a particular supplier kiln in Gujarat or Punjab to its final destination in the UK is in practice almost impossible for an outsider to the system.

In reality, it is often equally difficult for an insider, who in most cases deals only with a single distributor, without a wider oversight into the extended supply chain. Buyers of imported bricks can rarely offer the guarantees over the ethics of their

Consumer power in the global factory

product's production that their customers would prefer, so in place of transparency they offer the aesthetics of commercial distraction. Bricks labelled as originating from Gujarat, Punjab, or Lahore would inevitably attract a battery of questions from consumers, so they are instead given nostalgically, distractingly British names like Imperial Red, Rustica London Stock, or Suffolk Multi. By mirroring the names of domestic bricks, by alluding to tradition and domestic production, difficult questions are quietly sidestepped: a dirty trick, perhaps, but an exceedingly common one, with a long history far beyond the brick industry.

Greenwashing the global factory

A few years on from the Blood Bricks exhibition, I was on firmer ground and in more familiar circumstances. Once or twice a year, my department hosts an open day for prospective students, in which lecturers provide a taster of what they might expect from undergraduate Geography. My lecture looks at the hidden depths of global supply chains, and one of the tricks I enjoy including is to ask the audience where their shirt was made. After a few moments of fumbling with labels, at least a few of the shirts tend to say Bangladesh, or Cambodia, or another intermediary producer, at which point the rhetorical goal lies open. 'Yes', I say, 'it was sewn together in a factory there, but neither of those countries has a cotton industry, so where was it really produced?' It is a fairly standard, but satisfying, preamble to further information on the murky world of global production. It usually proceeds quite well, before a vaguely interested audience files out of the door and on to their next lecture. On occasion, though, a member of that audience might pause on the threshold, turn to me and ask 'so how do we *avoid* these things? What can we buy instead?'

Even years later, I feel the familiar rising panic. I know that I am expected to provide useful, tangible advice. I know that it is my obligation to provide answers to fit the problems I raise, but I cannot bring myself to offer definitive advice. So I string together something about second-hand clothes – usually received without much enthusiasm – and ethical consumer guides; my audience member leaves somewhat unsatisfied and I am left to reflect on why I find this question so difficult. After dozens of similar failures, this introspection has drawn me to an unsettling conclusion, one sufficiently heretical that it may well alienate a number of readers of this book. I do not believe that consumer power can produce a more ethical or sustainable global economy.

Allow me to explain myself. I do not, in the first instance, deny that consumers have power over products. There are countless cases of products being altered, replaced, or discontinued, and of new products emerging onto the market to satisfy consumer demand. The food industry is a good example of this. Changing consumer attitudes towards food and animal welfare have seen rapid increases in ethical food products in recent years. In the UK, the sales of food marked either organic, or with the Rainforest Alliance, Freedom Foods, Fairtrade, or another certification body has increased almost tenfold in two decades, from £1.3 billion in 2000, to 12 billion in 2020.[14] Across an even shorter and more recent timeframe, veganism is surging. Once a niche lifestyle, the number of vegans in the UK grew by some estimates more than four times in the five years from 2014 to 2018, from 150,000 to 600,000, or almost 1 per cent of the population. In the US, meatless and low-meat diets have taken

Consumer power in the global factory

an even firmer hold, with roughly 2 per cent of American adults considering themselves vegan, alongside 5 per cent identifying as vegetarians and 3 per cent as pescatarians.[15]

The food industry has had to work hard to keep up with these trends. Supply chains have become more flexible in order to respond to consumer demand, placing consumers in the driving seat of changes in food production and supply. Supply chains become more demand oriented and food production more directly responsive to signals from the market and feedback from consumers.[16] All of this is very heartening. It signals a growing interest in the ethics of consumption and production and a genuine intention to effect change in the food sector, even if it comes at personal expense and in some cases even the sacrifice of much-loved meals.

Here, though, is the crux of the matter: in order to achieve this flexibility, in order to respond more efficiently to the preferences of consumers, food supply chains have had to follow a similar trajectory to many other consumer products, enacting a long-term shift away from local, predominantly nationally centred food geographies, to complex international supply chains.[17] Despite a strong cultural ethos of sustainability, often linked to local, ethical production, the rise of veganism and its vegetarian and pescatarian sibling diets has not seen a return by food supply chains towards locality and seasonality. In most cases the opposite has happened. The rise of 'Big Vegan', heralded most recently by the arrival of KFC into the vegan marketplace, but also more generally by the rise in 'high-tech, ersatz, "ultra-processed" foods' designed to replace meat and other animal products signal this disjuncture most clearly, but it is one that extends far further, beyond food and into the wider geography of global production.[18]

Greenwashing the global factory

And it is not new either. The history of sustainability is littered with examples of companies hearing the desires of consumers, responding to them, and presenting ostensibly what the customer base wants. It is a practice that is as old as public interest in sustainability itself, but can be traced most clearly to the 1960s, when the first wave of environmentalism took hold of the public consciousness. Early efforts to cash in on the green dollar by making products seem sustainable were termed 'ecopornography' by the environmentalist and former Madison Avenue advertising executive Jerry Mander in the 1960s, a term later supplanted by the still common 'greenwashing' coined by Jay Westervelt in 1984. Originally used to describe hotel chains' use of signs asking customers to 'save our planet' by using towels more than once, it has stuck around as a catch-all term for the dubious environmental claims of corporate advertising.

There are countless similar examples. Even on the first Earth Day, on 22 April 1970, major corporations were already in on the act. Coca Cola proclaimed its 'bottle for the age of ecology': in reality, the exact same glass bottle as before, only now lauded for its recyclability. In the same year, Chevron, the petrol company, announced the launch of its F-310 petrol, marketed as producing 'significantly reduced air pollution from auto exhausts'.[19] Announced as 'the most long-awaited gasoline in history', it was accompanied by slogans such as 'cleaner air ... better milage' and images of balloons being filled by thick black smoke from a regular exhaust pipe, beside others pumped with clean, clear air from an exhaust pipe emitting F-310. It was a highly successful campaign, but it was also a complete fabrication. The 'revolutionary' active ingredient heralded by Chevron was already in use in other fuels

Consumer power in the global factory

and had no discernible impact on the emissions produced by a vehicle. As the Federal Trade Commission summarised in its complaint against Chevron, the product wasn't new and it didn't work.[20] Following an extended legal battle, Chevron agreed to discontinue F-310 in 1975, albeit after five highly lucrative years.

With these first shots already fired against corporate greenwashing more than half a century ago, the practice had to evolve – and it did. Greenwash advertisements flourished in the subsequent years, becoming more numerous and more sophisticated in the 1970s and 1980s. By the time Earth Day 20 rolled around in 1990, a quarter of all new household products that came on to the market in the US advertised themselves as 'recyclable', 'biodegradable', 'ozone friendly', or 'compostable', whilst 77 per cent of Americans said that a corporation's environmental reputation affected what they bought.[21] Environmental reputation was now a key strategic priority for many companies, but a slew of legal cases had begun to discourage the kind of disprovable claims made by Chevron twenty years earlier. Far better, instead, to achieve the much-desired green reputation through association, implication, and misdirection.

There is a huge array of such tricks still in use today, but one of the most popular is the simplest: merely to colour the packaging of the product in question green, add a few images of fruit or wildlife, and liberally apply the word 'nature'. It is an approach employed time and again, from 7-Up's 'Now 100% natural' slogan (to the extent that everything is *originally* natural), to Greenworks cleaning fluid (distinguished primarily for being green in a chromatic, rather than a sustainability sense), Mentos's 'freshly picked gum' (factory processing and an array of additives besides), and of course the ever-present Coca Cola

Greenwashing the global factory

Life (which uses stevia instead of sugar, a dubious green credential from any angle).

The only problem with this approach, as evidenced in particular by the failure of Coca Cola Life, is that everybody is up to it. Almost every product has green credentials of some description, even if only by implication. It is no longer sufficient to merely allude to a product's ethical qualities alone, so environmental claims have in a sense come full circle: bigger, more tangible claims are now needed, but under the far greater scrutiny of a professionalised environmental movement and well-established legal precedents concerning corporate assertions of sustainability and ethics. This is in many respects a sign of the success of the environmental movement. If they want to reap the benefits of the green dollar, companies do really need to do *something* to support the claims they make. They also need to ensure that nothing in their supply chain visibly contradicts their stated sustainability commitments.

'Visibly' being the operative word. Successes and failures like these support the idea that change is potentially possible, but also the need for a healthy degree of scepticism. Yet neither of these points is the root of my doubt over consumer power in global production. On the contrary, my problem is not so much about the power of consumers to *do* as their power to *know* a) what they need to do and b) whether their actions were effective. This is the weight tipping the scales of consumer power overwhelmingly in favour of corporate actors in recent decades. It is not only food supply chains that have expanded in their geographical scope since the first early forays into greenwashing in the 1960s. It is almost every supply chain. In the last half-century, the manufacture of everything from clothing to electronics to construction materials has been fundamentally

Consumer power in the global factory

reordered, exploded outwards onto an increasingly integrated, fast-moving global landscape of production.

This didn't emerge from nowhere. After all, as we saw in the previous chapter, the history of cotton production is deeply and often violently international. As with many products, the economic work of the colonial cotton industry was to globalise the three factors of production: land, labour, and capital, forcibly mobilising people and materials across different continents through the development of a militaristic logistical regime enforcing the enslavement of millions of people from West Africa, the shipping of millions of bales of cotton from the Southern US, and the investment of millions of pounds in industrial technology in Britain.[22] Earlier empires enacted similar disruptions in the service of the colonisers. The Roman economy was rooted in the extraction of people and goods from their origins: wheat from Sicily, Spain, Egypt, and North Africa, oil from Africa, wine from Spain, and slaves from everywhere.[23] And this is without mentioning history's most iconic supply chain: the silk road, which stretched thousands of miles across Eastern China via India, Persia, Arabia, Egypt, and Western Europe for fifteen hundred years, from 130 BCE, connecting centres of geopolitical power in Europe and Asia.[24]

Globalisation in the twenty-first century has largely settled into the grooves etched by these long centuries of imperial extraction. Yet, despite its historical foundations, this is an era in which new ground is also being broached. The supply lines of empires past are now supercharged by an unprecedented degree of connection, made possible by huge technological leaps in communication and logistics. Whereas international supply chains were once the arteries of trade, they are now the veins and capillaries too, reshaping distant environments to the whims of consumer countries. Production, as a result, is no longer local by default, because distance has been relegated several rungs below cost in industrial decision making by a series of major innovations, each of which progressively transformed the scale and efficiency with which rich nations are able to extract materials, process goods, and return waste to the global periphery.

The railways that underpinned the European empires of the nineteenth and twentieth centuries were amongst the most significant of these, allowing frictionless transport of goods across huge distances, but a far less heralded innovation, the standardised stackable container, was arguably of equal importance. This unremarkable-looking metal box transformed the world of trade, allowing complex journeys to be planned and paid at standard prices, pushed ever lower by mechanisation.[25] Rather than ships' holds filled to bursting with piles of goods of all descriptions, requiring skilled human labour to load, unload, and move onto onwards transport, these containers changed the logistical game. Simply lift them down with a crane from the open top of a container ship and onto the truck or train that will bring them to their destination. No need even to open the box along the way. It was the can that changed the world.

Even then, though, one final catalyst remained to kick start the current era of globalisation. Up until the 1970s, it was still not possible for a company to build a truly global factory, due to regulations that prevented or impeded foreign ownership in many major producer countries. If a company in the US wished to take advantage of the huge supply of low-cost labour available in China, for example, it would have to either outsource production (and with it profits) or invest capital in a facility they could not own. The deregulation of the global

Consumer power in the global factory

economy in the 1970s and 1980s, led on a world-altering scale by China, changed all this, paving the way for global supply chains in which single factories straddle thousands of miles, with freight logistics effectively playing the role that conveyor belts used to under the old domestic, Fordist model of production.[26]

The development of this system is important to understand because it shows us not only what happens in the course of contemporary industrial production, but more fundamentally the ways in which we *know* what is happening. As production has become increasingly, genuinely global in its reach, the goods and processes that constitute the basic necessities of our lives have had to become standardised and disaggregated into component parts in order to flow through the global factory. The production of goods under this global system depends on goods being categorised and recorded as they enter and exit containers, at which point they are no longer directly observable. Their path can be traced only through logs and manifests at either side of their journey. Which particular characteristics of those goods are recorded in those logs is the result of corporate priorities on the one hand, and systems of governance far removed from empirical realities on the other. We don't get to choose, so we don't get to see.

This fast-moving, standardised invisibility is a key obstacle to genuine accountability in supply chains. In a single, physical factory building it is possible to directly observe the materials and practices of goods being made. Production flows can be seen, touched, and inspected as they move; multiple processes can be directly observed simultaneously, but in the global factory this capacity is much reduced. The regular central oversight of the production process that is possible within a single space

Greenwashing the global factory

is replaced by remote monitoring arrangements, calibrated to precise technical indicators.[27] Visual and physical connection are replaced by check boxes.

Under this worldwide system of production, inspections of supply chain processes remain possible, but not regularly possible. The physical logistics – and to an equal extent, the cost – of such inspections means that it is far too expensive to regularly inspect the various 'departments' of the global factory. It would simply not be feasible, in most cases, for an international brand to send out emissaries to each stop along a long and complex supply chain on a regular enough basis to achieve meaningful oversight. Periodic visits may well be undertaken, but this is far from the same thing. Any practice can be hidden, or swept under the carpet for a day, or even for an hour, as the case may be, so faith is placed instead in intermediaries, who inspect on behalf of a brand or buyer and guarantee the conditions under which the product sold is made. There is no reason, in principle, why this cannot work, but in practice the remoteness of external supervision provides ample opportunity for deviation from the standards claimed by brands themselves.[28] Despite the power of technology, the obscurity of distance still shrouds what is really happening in a thick fog of uncertainty.

Lost knowledge in the global factory

Although I always feel bad not offering better consumer advice, each visit to the coalface of production deepens my scepticism of easy answers. The conversations I had in early 2022, a few months after I had failed to enlighten the last group of visiting students to my department, did little to dispel this conviction. I was in the central Cambodian province of Kampong Speu,

Consumer power in the global factory

sitting opposite a local union representative of a factory producing garments for several major international brands. We had arranged to meet in a roadside noodle shop, to speak about environmental issues that factory workers were facing. Interviews like these can sometimes be slow going, as workers can be afraid of saying the wrong thing, or unsure of how best to articulate their perspectives, but in this case the representative got straight to the point. As a large factory, producing for big names in the US and UK, this was precisely the kind of workplace that *does* get inspected. Yet, far from providing examples of good practice, Vishal the unionist's descriptions of these visits shed a harsh light on oversight in international supply chains. As we sipped iced tea, he observed the following:

> When we meet the brand, the factory sets the lines for us to do the presentation. We cannot just say this and that. They said not to make any issues for the factory. For example, if they asked about the temperature in the factory, we had to say it is not hot and we cannot tell the brand team that we work at the weekend.

An hour or so later, I left the noodle shop, still mulling Vishal's candid complaint whilst heading for a second factory in the local area. My second meeting of the evening was with a worker called Sokhorn, in the forecourt of a shuttered roadside grocery shop. As the traffic zoomed past on the national highway just a few metres away, stirring up a heavy mix of dust and fumes into the warm night air as we spoke, Sokhorn discussed the difficulties of working in the factory, the heat that takes workers days at a time to acclimatise to, the heavy workload, and the physical strain of working long hours in these conditions. At the root of his complaint was a faith that if the brands only knew what it was like, then they would surely improve matters. Once again, the now familiar account of ineffectual monitoring came up, this time more explicitly still:

> The CEO of the brand doesn't come to the factory. They have assistants who are Cambodian. So, if there were some issues, their assistants can hide them from the brand owners. We don't know whether they get the commissions from the factory or not, but the garment workers always feel pressured because of their work. Some places command the workers to work in unsafe working places. They carry heavy loads. Some sectors are not safe, so they kind of risk their life as well.

As I drove back that evening, I considered the implications of what I had heard over the past few hours. I was well used to stories of the difficulties faced by workers in the garment industry, but my questions that day had been predominantly environmental. I had set out to understand whether workers were affected by the increasing incidence of extreme heat that Cambodia had been experiencing in recent years, so I hadn't expected the workers I was talking to to bring up these problems of observability so often, or so forcefully. Before long, I reasoned that I should not have been surprised. The politics of the supply chain runs through all aspects of garment workers' lives and they are naturally aware of it, especially as it is commonly used against them.

When floods force the closure of factories, for example, as they did in Phnom Penh in late 2021, workers' salaries are cut until work can resume: sometimes by half, sometimes by 80 per cent or even 100 per cent. Factory bosses explain this to workers by saying that they are paid on delivery, so if clothing can't be made, then no money is made. This much is true, but the more important point is how responsibility is passed up and down the supply chain. Brands do not accept direct

Consumer power in the global factory

responsibility for workers, agreeing contractual arrangements where factories are offered no support in difficult periods. The arrival of Covid-19 provided an even bigger example of this. Faced with a sudden drop in demand, many brands refused to purchase goods they had ordered and which factories had already produced, ready for shipping.[29] Partly as a result of this, workers lost their jobs in their tens of thousands during the pandemic,[30] but brands view this as entirely beyond their responsibility. Factories do not belong to brands and brands do not bear legal responsibility for what happens inside them.

This is not to say that brands don't care about workers. In fact, I know a number of people who work for major brands who are both aware and concerned about these issues. The issue is rather one of accountability. As industry has expanded into the present global factory, it has effectively outpaced the legal frameworks intended to regulate work. The rights held by workers in the UK, Europe, or even the US were hard won over decades, even centuries, of struggle. In many cases, it took world wars to catalyse the development of statutes on pensions and working hours.[31] Yet as production has moved increasingly beyond the borders which delimit those legal frameworks, they have begun to lose their universality. Today's global supply chains transect a patchwork of different environmental and labour standards, in which the legal frameworks and their enforcement differ markedly. Rather than all workers within the global factory having the same system of rights and obligations, each department effectively has different ones.

Instead of legal frameworks, consistency in supply chains is achieved instead via corporate social responsibility standards in which the large-scale firms voluntarily agree to regulate labour and environmental conditions in their supply chains.

69

Greenwashing the global factory

Over 90 per cent of Fortune 250 firms in the US have signed up to some form of these agreements.[32] How effective arrangements like these are, however, is highly questionable, as exemplified in a study of Gap's supply chain between 2010 and 2016, which found no correlation between the imposition of voluntary frameworks and subsequent improvement in standards. When penalties were subsequently applied for non-compliance, a relationship did begin to appear, but it remained marginal, with non-compliant factories only 22 per cent less likely to fail in future assessments.[33]

And this is before we even consider the real booby trap of sustainable production. Even if factories are made to comply with high labour and environmental standards – and there are many examples where this has been successfully achieved – there is no guarantee that the goods a buyer ends up with were made, or at least wholly made, in that factory. In the garment sector, this is a huge issue. A recent survey of the sector found that some 36 per cent of all orders were subcontracted without authorisation, but in Cambodia the figure is higher still, with almost 55 per cent of all orders subcontracted.[34] This vast shadow industry of hundreds of small-scale factories is almost completely unregulated. It is low paid, workers have no employment security to speak of, and conditions are notoriously tough. Many subcontracting factories are simply tin-roofed roadside shacks, in which several dozen workers sweat the day away adding a single detail to a batch of garments: perhaps a certain type of button, a frill, or a 'Made in Cambodia' label.

These workers, toiling over the very same garments that will soon grace the shops of major brands in the global North, are effectively invisible. The minimum wage rises of recent years

Consumer power in the global factory

have passed them by; corporate social responsibility and best practice supply chain standards are irrelevant. As far as brands are concerned, the silent workforce that produces more than a third of the orders they receive does not exist, but this is no more a garment sector problem than it is a Cambodia problem. It is a symptom of the ways in which we structure global production at the present time, with large-scale buyers able to call the logistical shots and producers having little option but to comply. When failing to complete an order may result in the loss of both the capital invested in the attempt and potentially the relationship with the brand that ordered it, failure is not an option.

As a company, it is not OK to say this of course. You can't tell a US consumer that you think you have a vast shadow workforce, or that you have only limited knowledge of what goes on in the factories making the clothes you're selling. So the apparent objectivity of the systems set up to manage supply chains is crucial to keeping the whole business running. Most of the time this works well. A brand sets up some indicators to monitor, those targets are shown to be met at periodic intervals, box ticked. Only when a new issue emerges that isn't included in these compliance frameworks do the limitations of this system of remote monitoring come to light.

Having recently completed a study into illegally harvested forest wood usage in the Cambodian garment industry, I know this from personal experience. It is an endemic practice that had been rumbling on for years as a half-known issue. I was first alerted to it back in 2013, when a journalist showed me grainy phone photos of a large truck rolling into a garment factory, but I didn't give it my full attention at the time because at that point I didn't fully understand why it would be used, or how widely. As it turns out, it's for ironing. Electricity in Cambodia is amongst the most expensive in the world and it is not cost effective to use it to generate the large amounts of steam needed every day in some factories. So, in large ironing depots, hundreds of tons of forest wood are brought in by truck every day, filling vast depositories stretching hundreds of metres. Since burning forest wood became illegal in 2018, these trucks now come at night, lining up one after the other in the early hours, but gone before daybreak.

Shocking as it might seem, this is no secret. On the contrary, it is widely known. Local communities know, factories know, their workers know, and brands know. At least 30 per cent of factories in Cambodia are engaged in this practice.[35] Yet as it doesn't figure in official reporting, it doesn't exist to consumers, who are presented with the usual 'zero deforestation' pledges by companies that burn forest wood to make clothes. The secrets of the garment industry, alongside many others, are hidden behind high walls in remote parts of distant countries. To work out what's going on in the absence of transparency you have to find a way to approach those walls and peer over them. It requires travelling hours in a car, snooping around factories, peering through gaps in fences, speaking to local people. In some cases, it might mean flying drones above suspect factories to get a new perspective. Any one of these things is far beyond what an engaged ethical consumer could do, so with so much of the global economy shuttered thousands of miles away from the people who buy its products, it presents an effectively insurmountable barrier. The contemporary reality of supply chains is that they are so knotted and impenetrable that the very companies that use them cannot elucidate, let alone reshape them to their own ends. What hope, then, the consumer, of peering

Consumer power in the global factory

successfully through the murk? However good the intentions, the reality is often barely a glimmer.

And this, at base, is the reality that consumers face when seeking out an ethical purchase. Removed from the direct political governance of national production, manufacturing in the global factory is effectively a black hole: a twenty-first-century consumer incarnation of the age-old 'terra nullius' long exploited by rich nations. Companies enact standards on their supply chains, yes, but these standards are self-defined and self-enforced. Without independent oversight and scrutiny, global corporations are effectively free to make any claim they wish; naturally, a situation that suits them. A green image is highly lucrative because consumers really do want green products, so without having to worry about the veracity of their claims, global corporations are able to devote their attention to publicising them. This is the illusion of green capitalism.[36] On the high streets of the rich world, there is barely a product on sale today that does not make green claims of some sort. Yet, in the absence of verification, these claims are merely a lucrative illusion: greenwashing at best, outright lies at worst.

So, for those consumers who want to know what to do about environmental abuses overseas, herein lies the answer. It is not, unfortunately, as simple as making a purchase, but it is vital and it is possible. Rather than seeking out green products, consumers must use their voice to lobby for oversight of those products: an independent authority to ensure supply chains are as clean as they appear. For a long time this has been unthinkable, but the first green shoots of change are sprouting. In January 2022, the German parliament passed the Supply Chain Act,[37] a law that opens the door for independent oversight of global supply chains based in Germany and action

Greenwashing the global factory

against corporate offenders. It is far from perfect and critics brand it a political compromise, but it marks the beginning of a paradigm shift away from corporate self-governance of the supply chains that underpin our lives.

If you are a consumer who doesn't live in Germany then this is what you can do about it. Make your voice heard, so that globalised abuses can be seen. The next time you find yourself researching which brand of t-shirt, coffee, or petrol is the greenest, save your time, go home, and write a letter to your MP, congressperson, or député demanding a rigorous supply chain law. Your letter on its own will make no difference, because individual voices are all too easily drowned out, but with enough of us involved change is possible. Each time you return from the supermarket, convince one more friend or relative to send a letter and eventually the weight of public opinion will tell. If you can spare the time, get involved in local politics; join your council or local political party and encourage as many people as you can to make their voice heard.

This is not easy, because before you can begin to build a movement for change you have first to wean people off the opium of sustainable consumption. Do not discourage or disempower, but redirect the energies of the people you encounter towards politics and legislation. It may seem hardhearted, but taking away the moral pressure valve of the ethical purchase is imperative. The majority of people are deeply concerned about climate change, but their efforts are channelled fruitlessly down blind alleys. Each green decision, each eco-conscious choice, should instead be a demand for scrutiny, justice, and change. Demand a light be shone into the dark corners of our global supply chains. Put down your ethical consumer guide and pick up your pen and your phone. And when people ask you how

Consumer power in the global factory

they can avoid the worst aspects of the global economy, when they ask what they can *do* about them, tell them plainly. As individuals, nothing, but collectively, we can demand an end not to one abuse, but many, by taking back control of our economy, our production, our climate.

4

Carbon colonialism: hidden emissions in the global periphery

After returning from Cambodia, I spent the summer of 2019 as a fellow of Humboldt University in Berlin, an experience that helped enlighten me to the shortcomings of well-intentioned sustainability thinking. Whilst meeting some friends at a café one Sunday the topic turned to climate change and what could be done about it. It is the kind of extended, ethical conversation that often feels like the mainstay of café culture in parts of Berlin, frequently overheard above the clinking of cutlery and the whirring of bicycles over cobbles. Knowing that I worked on the topic, I was asked what lifestyle changes I would recommend, to which I gave the usual answers. 'And so, if we can persuade everybody to do this, then everything will be OK, right?' 'No.' 'But I don't understand, if everybody cuts down their impacts a little bit then overall it must decline?' 'Unfortunately not. It's not enough.'

Once again, I was being difficult. Berlin is an enthusiastic adopter and believer in sustainable consumption. Organic products are widely available and clearly labelled with a universally understood 'bio' symbol, everybody reuses their shopping bags, and you'll almost never see a plastic straw. It's such a part of the city's identity that the Berlin marketing organisation visitBerlin introduces the city like this:

> Berliners love to 'go green'. Urban farming, green fashion and vegan gastronomy are turning the former industrial city of Berlin into an ever-increasing green metropolis. Change is driven by a desire to preserve and expand green oases, coupled with the willingness to follow sustainable and alternative paths.[1]

'Going green', in other words, is fashionable. It's modern. It's aspirational. It's something the vast majority of people want to do, not just in Berlin, but everywhere. In the UK in 2021, for example, 82 per cent of people agreed that if everyone did their bit, we could reduce the effects of climate change, according to the Office for National Statistics.[2] Reflecting this positive attitude, 83 per cent of people reported at least one environmentally friendly adaptation to their lifestyle in March 2020, with the most common being avoiding or minimising food waste. Figures like these are typical of many countries. They highlight the serious concerns that most people now have about climate change and the genuine will that most people have to tackle it, even if it costs them more, takes them more time, or means their making compromises to their lifestyle. They also highlight how completely intertwined the ideas of consumption and sustainability have become in recent years. The idea that we can consume our way out of the climate crisis is implicitly accepted by most people, as is the idea that small-scale actions aggregate to global effects.

This is because climate change discourse over the last two decades has been dominated by what is known as the 'one world' narrative, the culmination of a wider trend beginning after World War II, whereby the implicit meaning of the word

Greenwashing the global factory

'environment' has expanded from meaning a person's immediate environs – for example, their street or their village – to
its current interpretation as the global environment.[3] This has
meant a profound shift in the scale at which we think about the
world, rooted in the rapid expansion of transport and communication technology, combined with key cultural moments like
the iconic Earthrise shot of the Earth from the Moon, taken by
astronauts during the Apollo space programme.[4] Over time,
the idea of the global environment has become so commonplace that it has reshaped how we view our relationship with
the planet.

To exemplify this, let's consider a recent intervention on
the topic of climate change by the big budget US *Late Late
Show* with James Corden. Entitled 'We're All in the Climate
Fight Together', the show comprised Corden and his co-hosts
addressing 'some positive, inspirational news stories about how
people and companies are stepping up and rising to the occasion to confront climate change'. The small things that everybody could do included buying a water-saving shower head
and not replacing an old Prius. There are many, many similar
examples, but this one highlights perfectly how far scale has
been devalued when it comes to climate change. The idea that
small actions by (predominantly wealthy) individuals contribute
in a meaningful way to the global climate, much as consumer
attitudes might see a rise or fall in the sales of a given product,
is hard wired into public discourse on climate change. What's
more, this 'one world' framing is often the conduit through
which the best of intentions are expressed. As the closing line
of a commentary entitled 'On Climate Change, We're All in
this Together' in the *Virginia Mercury* newspaper concludes, for
example, 'there are no "other" people we can exclude, neglect

Hidden emissions in the global periphery

or destroy to save ourselves. There is only all of us, all across the world, and we are all in this together.'[5]

And this is far from all. From Pope Francis's *Laudato Si* declaration on the environment,[6] which refers to the climate as the 'common good' of 'our common home', to the UN Secretary General Antonio Guterres's declaration at the One Planet Summit in 2021 that 'we are destroying our planet, abusing it as if we had a spare one',[7] the one world message is everywhere: 'we', meaning everyone in the world, are destroying 'our home', meaning the whole world. The underlying implication, that our responsibility is equal, our vulnerability is equal, and, crucially, our capacity to act is also equal, is a hugely powerful force in the landscape of sustainability. Yet it is grossly misleading. In fact, the idea that the global aggregation of small-scale collective action is the key to 'solving' climate change is perhaps the single biggest roadblock to effective action on climate breakdown.

So, let's put some of the little things we can all do into perspective. One of the main changes that individuals are advised to make to live sustainably is to fly less. Flying is the bête noir of sustainability and on an individual level it does indeed comprise a massive chunk of a global Northern person's carbon footprint. For the average American, flying contributes about 583 kg of CO_2 each year, slightly lower than the average British person at 840 kg, and substantially lower than the average inhabitant of the UAE, who averages almost 2 tons of annual CO_2 from aviation alone. This sounds pretty big. And indeed it is, especially when you consider that the *total* annual carbon footprint of a person from Ethiopia, Uganda, or Nigeria is only 100 kg, but it looks rather different in the light of the equivalent annual figures for the US (15.97 tons), UK (5.46 tons), or the

Greenwashing the global factory

UAE (15.78 tons). Flying constitutes 3.7 per cent of the average US carbon footprint, 15.4 per cent of the average British one, and 12.7 per cent of the average UAE one. And this is only taking into account the domestic carbon footprints of each country, something we shouldn't do for big importers like this. If we think in terms of what the average citizen consumes, the figures fall to 3.4 per cent in the US, 10.92 per cent in the, UK and 9.69 per cent in the UAE.[8]

So even for the countries that do it most, flying is not perhaps as big a deal as it might seem. At a global scale, this is even truer. Aviation is small fry in the grand context of planetary carbon, contributing only 1.7 per cent of global emissions,[9] a figure dwarfed by livestock and manure (5.8 per cent), agriculture, forestry, and land use (18.4 per cent), and industrial energy use (24.2 per cent). Far worse than this, it is a figure dwarfed even by the rate of global carbon emissions *growth*, which averaged 2.2 per cent each year between 2000 and 2020.[10] To put it bluntly, if every person in the world became so suddenly and acutely aware of their own responsibility to halt climate change that they foreswore flying completely and indefinitely on 1 January next year, then all else being equal there would *still* be more carbon in the atmosphere on the following New Year's Day.

This should not, perhaps, be as shocking as it seems. After all, the Paris agreement of 2016 was only the latest in a long-running series of international agreements to make limited tangible impact on emissions. From the first World Climate Conference in 1979, via the UN Framework Convention on Climate Change in 1992, Kyoto in 1999, Copenhagen in 2009, and finally Paris in 2016, agreements have become more specific and binding over time. Yet all the while the atmospheric

Hidden emissions in the global periphery

CO_2 concentrations have continued to increase. At the time of the first World Climate Conference, atmospheric CO_2 stood at 339 parts per million; at the foundation of the UNFCCC 13 years later, it was 358. As the fireworks boomed in Paris it was 402 parts per million; and today, it stands at 421.[11]

The apparent lack of impact achieved by these high-profile agreements presents something of a conundrum to environmentalists. Each meeting has, to a greater or lesser extent, agreed frameworks and policies with the world's heaviest-emitting nations that would be expected to reduce carbon emissions. Moreover, the data show that in many cases, they *have* resulted in reduced emissions. The EU's net emissions fell from 5.6 billion tons of CO_2 in 1990 to 4.2 billion in 2018,[12] whilst the UK – historically one of the EU's largest emitters – claims a 44 per cent reduction in emissions since 1990.[13] Even the United States, a country whose efforts have been deemed 'critically insufficient' by monitors, has achieved a modest decline, from 7.1 billion tons in 1998 to 6.7 billion today. China, one of the world's largest and most rapidly increasing carbon emitters of recent decades, has begun to slow the rate of increase, with emissions projected to plateau over the next five years[14] as part of a national plan to achieve carbon neutrality by 2060.

So what, then, lies behind this discrepancy? Emissions from major economies are either falling or stabilising, yet the relentless uptick of global carbon emissions continues undiminished. Are major emitters being untruthful about their emissions figures? Not in a direct sense. Rather, it is the reductions themselves that are illusory, the product of a system of carbon accounting which remains firmly national and bordered within an increasingly global and interconnected world. To put it simply, as richer nations increasingly diminish their share of global industry, 'outsourcing' lower margin and more environmentally damaging processes to the global South,[15] the emissions associated with those processes – at least in headline figures – go with them.

In total, imported emissions – emissions that occur in a country other than where they are used – now account for a quarter of global $CO2$ emissions,[16] a figure that highlights the scale of wealthy nations' ability to move emissions off their environmental books.[17] There is even a name for this practice. The ability to effectively outsource emissions from richer to poorer nations has been described as 'carbon colonialism':[18] a term emphasising the historical power relations that underpin carbon accounting. Wealthier countries, overwhelmingly responsible for climate change both historically and currently, have set the terms of carbon mitigation at the negotiating table. Naturally, these terms favour the biggest emitters, allowing larger economies to offshore production processes to smaller ones, whilst maintaining the economic fruits of that production.

The scale of this outsourcing is enormous, in some cases completely undermining national claims to environmental progress. Take the UK for example. This is a country that makes great hay of its global green leadership, as exemplified by the ever-present figure of 44 per cent carbon emissions reductions since 1990. Yet, if we look at the bigger picture, most of these gains disappear. The value of UK imports has more than doubled in the last two decades,[19] with environmentally regulated EU exporters accounting for a falling proportion of the total.[20] With so much of British production now happening outside our borders, more and more of the carbon involved in making what British people use every day is being added to the carbon budgets of other countries, rather than that of Britain itself.

Hidden emissions in the global periphery

Britain is now the G7's largest proportional importer of emissions, with carbon consumption from imports now 28 per cent higher than 1997 in absolute terms.[21] This rise in imported emissions chips away substantially at the UK's much-trumpeted domestic emissions reductions, reducing it from the government's gross 44 per cent figure to a net 15 per cent reduction.[22] Rather than the substantial reductions claimed by the UK government, therefore, the last two decades have seen a concerted shifting of emissions onto foreign territory.

On a planetary level, this is clearly a problem. A temperature rise of more than 1.5°C will likely result in 'several hundred million' more people in poverty by 2050[23] and the lack of progress in slowing global emissions makes this practically inevitable. Yet, beyond these global figures, there is also a smaller-scale human and environmental cost. Removed from the regulations and standards governing domestic production, the industrial processes that manufacture the goods consumed in global Northern countries are often dangerous and environmentally destructive. By outsourcing these impacts into long and complex supply chains, wealthy nations are able to hide the damage their productive processes engender, not by resolving it, but by moving it across a national border and thus largely out of sight of regulation and accounting.

What this system means, in a nutshell, is that climate change impacts, including the slow-burn disasters of droughts and floods, are being effectively traded out by wealthier countries and imported by less wealthy ones as the price of economic growth. All the while, this environmental degradation remains hidden by the analytical legacy of nationalism, an emphasis on the structures and strictures of the nation state no longer appropriate for a globalised and interconnected world.

Clothing the garment industry's carbon footprint

In many ways, we already know this. After all, of all the sustainability trends in green-loving Berlin, one of the most immediately visible is the city's love of vintage clothing. In the fashionable Southeastern districts of the city, it feels as if there is a vintage clothing store on almost every street, from the large-scale chain Humana to the dozens of smaller boutiques that sell specially selected pieces at prices similar to new garments. And Berlin is far from unique in this. You can find similar densities of second-hand apparel in London, New York, Barcelona, Naples, to name but a few. There is often nothing said about sustainability in these boutiques because it doesn't need to be. From sweatshops, to emissions, the ethical shortcomings of the apparel industry are widely known.

But how well do we really understand an industry so complex that, in a typical wealthy country like Germany, the five biggest manufacturing importers – C&A, H&M, Zalando, Inditex, and About You – employ 5,235 factories in 60 different countries?[24] Certainly, there are some things we do know. The garment industry is acknowledged to be a major contributor to climate change, accounting for between 5[25] and 10 per cent of global emissions.[26] Yet, despite a growing awareness of 'the price of fast fashion',[27] the scale and complexity of the industry, as well as the opacity of supply chains defined and delimited predominantly by large-scale end-stage buyers,[28] means that much of its impact continues to be underestimated.

What is perhaps more surprising though is that even individuals with substantial influence over the garment sector, indeed even brands themselves in some cases, face similar issues in practice: an issue exemplified by recent humanitarian

Hidden emissions in the global periphery

concerns over the Chinese cotton-producing region of Xinjiang. Following a wave of reports of forced labour in the region, numerous brands sought to divest from Xinjiang cotton, citing 'grave concerns about systemic, state-sponsored coercion' in the annual cotton harvest.[29] By the time the Better Cotton Initiative had removed its seal of approval from Xinjiang,[30] several brands, including Uniqlo, Calvin Klein, and C&A, had issued statements claiming not to source any materials from the region. Yet the complexity of the apparel supply chain made this an almost impossible claim to back up in practice.

China is the world's largest producer of cotton, accounting for some 20 per cent of world output, within which the Xinjiang region produces the lion's share of national output, at 84 per cent.[31] This substantial share means it is difficult to avoid, especially in Southeast Asian intermediary manufacturers, which are heavily dependent on cotton imports particularly from China. In Cambodia, as in neighbouring Vietnam, there is almost no domestic cotton industry at all, meaning that the raw materials for cotton-based garment production must be imported from overseas. Of the cotton imported to Cambodia, for example, 81.6 per cent arrives directly from China and a further 4.8 per cent via Hong Kong; meaning that a total of 86.4 per cent of the cotton used in Cambodia originates directly or indirectly from China. Assuming that Chinese cotton exports are produced and imported in the same proportions as the national average, this suggests that some 71.5 per cent of cotton garments made in Cambodia, including those exported to the UK, are made with cotton grown in the controversial province of Xinjiang.[32]

These raw figures suggest that it would be difficult, if not impossible, for all of the UK's major apparel companies to be honouring their ethical commitments. Yet, as brands themselves concede, the obscurity of extended supply chains makes it, in the words of Stella McCartney herself, 'extremely difficult'[33] to prove or disprove linkages between end retailers in countries such as the UK and regions such as Xinjiang. So even in a case as high profile as this, concerning ethical infringements characterised by the US government amongst others as 'committing crimes against humanity and possibly genocide against the Uyghur population',[34] the garment supply chain is simply too knotted to disentangle.

And there's more. In addition to its ethical implications, supply chain obscurity of this sort makes hiding carbon costs exceedingly easy by making the origins of materials very difficult to trace. Once again, China is at the centre of this story. Between 2014 and 2018 almost a third of China's total cotton supply was imported, making it the largest importer of cotton in the world, as well as the largest producer. Of this third, the vast majority of imports came from five countries: Australia (25.9 per cent), the USA (28.9 per cent), Uzbekistan (7.75 per cent), India (12.9 per cent) and Brazil (12.7 per cent). Besides a small contribution from Mexico, the remaining 7.5 per cent of imports were produced predominantly in Africa, most notably Burkina Faso, Cameroon, Mali, Cote d'Ivoire, Benin, and Zimbabwe.[35]

Now remember that China processes cotton for almost all of Southeast Asia. This means that by the time your roll of cotton fabric even makes it into a factory in Cambodia, Vietnam, or Laos, it has on *average* already travelled 14,000 km to get there, emitting 63 g of carbon from transport even before the finished product begins the 18,000 km journey to the British high street. So, by the time you pick up your cotton t-shirt in a British high

Hidden emissions in the global periphery

street shop, it has already travelled 32,000 km, three-quarters of the way around the world and accompanied by total emissions from transport of 127 g of carbon per shirt. Applied to every garment that makes its way from Cambodia to the UK, this adds up to 30,867 tonnes of carbon annually, of which almost half – 12,612 tonnes CO_2 – comes from that long, raw material supply chain. That's the equivalent of 167 tanker trucks full of petrol being burnt, or the average total annual emissions of 4,264 Europeans, concealed by the logistics of global production.

Now imagine these numbers scaled up to truly reflect every product sold globally. Figures like these illuminate the otherwise invisible systems underlying our everyday lives, casting doubt on many of the assumptions we make about sustainability. The lack of transparency surrounding global supply chains means that many sources of emissions are either hidden or significantly underestimated. And their extraordinary complexity impedes detailed analysis and undermines accountability, concealing many carbon emissions from public view.

When you pick up a t-shirt in a high street shop, it might say made in China, Bangladesh, India, Cambodia, Vietnam, or one of many others, but in most cases, the single-country origin label sewn into your t-shirt is an illusion, reflecting a problem that affects so many of the items we purchase and use daily. In fact, that country of origin is just one stop on a global journey of assembly that is anathema to truly sustainable production and a key obstacle in our fight against the climate crisis. Understanding this hidden geography is the first step towards tackling the opaque and misunderstood carbon footprints of our global economy and decolonising systems of environmental accounting that favour the world's biggest polluters.

Constructing climate change

The garment industry, though, is something of a known unknown: we are aware of the limited information we have and aware that it's a problem. Far more problematic are the unknown unknowns: massive sources of carbon we don't even know we're not counting because they are lurking, often in plain sight, but far removed from official statistics. Not long before I arrived in Berlin, I had been made aware of just such a hidden source of carbon, when I had paid a visit to the head of an organisation representing British brick manufacturers. He turned out to be something of a larger-than-life character, sporting a white-collared striped shirt made famous by British bankers of the 1980s, union jack cufflinks, and an apparently deep well of support for the left-wing Labour leader Jeremy Corbyn.

It was an unusual combination, but not as intriguing as what he had to say on the state of the brick industry which, he told me, had contracted in the wake of the 2008 financial crash and could not now meet demand. According to HMRC records, he explained, a growing proportion of bricks – about one in seven – is now imported from overseas, with an ever-larger proportion of these originating from as far afield as India and Pakistan.[36] And so shattered an illusion I didn't know I had: that the very matter that surrounds us, the earth itself reshaped to fit the fashions of modernity, must, to some degree at least, be local. It was a recognition that seemed to me to transcend the importance of a single commodity, speaking instead to the fundamentals of the way we live. Bricks are not simply a part of our world, after all. We are living, for the first time in history, in an urban world, in which more than half of the world's population

Hidden emissions in the global periphery

is now living in towns and cities.[37] For the majority of us on this planet, bricks are our world: the inevitable, often invisible containers within which our lives are played out.

This has significant implications for sustainability. Urban areas and their construction are increasingly recognised as major contributors to climate change, with the built environment currently responsible for 39 per cent of energy-related CO_2 emissions worldwide.[38] Some 20 per cent of global black carbon is attributable specifically to brick kilns, 90 per cent of which are in Central Asia.[39] Not only is it a dangerous local pollutant, highly damaging to human and environmental health, but – despite its absence from most greenhouse gas reporting – it is also considered to have a significant effect on global warming.[40]

This is increasingly an issue that resounds beyond Asia. As many countries in the rich world transition away from manufacturing, their urban infrastructure is becoming ever more prone to carbon embodied in the burgeoning global trade in materials. The UK, for example, is now the world's sixth-biggest importer of raw materials,[41] highlighting the pressing relevance of the issue. Yet many elements of this wider trend have gone largely unnoticed. The UK imports over 30 million bricks each year from India and Pakistan alone, yet the environmental and social implications of engaging with an overseas industry recognised as a source of key humanitarian and socioeconomic issues, including modern slavery,[42] has garnered almost no attention in UK policy.

From an emissions perspective, this hidden trade is also a major issue, contributing vast volumes of carbon as bricks are moved from South Asia to Western Europe, a 17,000 km journey. Each 40 foot container of bricks making its way to the

UK emits an estimated 620 tons of carbon on its journey from South Asia. Combined with higher levels of emissions released in production compared with UK brick production processes, this means that non-EU bricks imported to the UK carry average embodied emissions 2.5 times more than UK bricks, or an excess carbon cost of 0.66 kg per brick. A standard house built with 8,000 of these bricks would therefore 'cost' 9,000 kg of CO2 emissions, equivalent to a car driving 23,000 miles. The excess carbon cost alone (i.e. compared with the equivalent house built with domestically produced bricks) would be 5,280 kg: over 13,000 vehicle miles, or 12 barrels of oil burnt.

In the context of the pressing need to reduce the UK's carbon footprint, these figures are clearly cause for concern. Yet what keeps the bricks coming is that the price is far lower. Bricks sourced in India cost on average £54.75,[43] a tiny fraction of the £686 charged for the same number of bricks in the UK. Even when factoring in the cost of transporting those bricks, roughly £39.51 based on a full 40 foot container of bricks,[44] the financial incentive to import bricks remains substantial. As a result, brick imports are continuing to grow rapidly. Between 2015 and 2019, the last available pre-Covid figures, the number of bricks imported from outside the EU increased almost tenfold, from 3,088,902 to 32,942,280.[45]

Despite the enormous scale, though, as far as environmental regulation is concerned this burgeoning trade is invisible. UK environmental legislation does not currently require even a disclosure of the emissions and environmental impacts associated with corporate supply chains, a gap in legal coverage that exemplifies how bulk imports of this sort fall through the cracks of environmental governance. Brick kilns are recognised as one of the world's largest stationary sources of black carbon and

Hidden emissions in the global periphery

are 'stringently' regulated by Defra within the UK's borders.[46] They remain the top air polluter in Bangladesh[47] and a key environmental challenge in numerous other South Asian brick exporters.[48] Yet the international, displaced nature of the environmental impact means importing governments don't view regulation as necessary.

This forms a stark contrast to the stringent environmental regulation applied to brick production within the UK's borders, as I was able to observe during a tour of such a kiln not long before leaving for Berlin. It was a sizeable operation, involving multiple large machines, some smaller artisan workshops, show rooms, and offices. Beside the kiln we were shown an area of land cut into the shape of a neat V, in order to remove the soil. We were told of the regulations governing the angle of that V, its size and location in the area. The work still looked hard, it still ultimately involved shifting large amounts of heavy matter from one place to another, but people were only one part of a wider operation in which machinery did most of the heavy lifting and strict legislation set the terms of work. This well-established system of rigorous regulation cut a jarring contrast against the complete absence of environmental standards applied to imported bricks. Clearly, it is not about the brick itself, nor even the emissions it generates, but about how environmental laws apply only to fragments of supply chains in a globalised, warming world where this kind of narrow focus no longer makes sense.

And the UK is far from alone in this problem. In the EU, the Energy Performance of Buildings Directive (2002) has driven building regulations in all member states to require increasing operational energy efficiency performance from 'regulated' energy, including fixed lighting, and space and water heating and cooling. Yet embodied emissions from construction materials was and continues to be excluded from the Directive. Indeed, whilst the European Union, on paper, achieved its goal of a 20 per cent reduction in emissions between 1990 and 2020, if we include a full lifecycle accounting of European member state carbon emissions, that is, one that includes consumption of imported goods like bricks, we get a completely different picture. When all of this is taken into account, EU emissions have actually grown by 11 per cent, with some nations seeing substantially higher emissions growth even than this.[49]

Far from ensuring a meaningful reduction in emissions, then, these examples highlight the way that domestic emissions accounting by wealthy import-dependent economies actively incentivises the outsourcing of emissions overseas. Referred to in technical terms as 'displacement',[50] or 'carbon leakage',[51] this is a widespread phenomenon that represents a major obstacle to the efficacy of environmental regulation. Yet it is of course no accident. Unequal trading relations of this sort have been a feature of the global economy for hundreds of years. And so they remain, the legacy of past injustices recast as diverging environmental vulnerabilities.

Carbon colonialism

Walking the streets of Berlin means being regularly confronted with the legacy of the past. Be it the small, brass plaques, hidden amongst the pavement cobbles, that mark the arrest of Jewish citizens, or Humboldt University's Empty Library that marks the twenty thousand books burnt under the Nazi Regime. It is a full-frontal engagement with a history deemed shameful to contemporary society and it stands out not only for its directness, but its uniqueness. After all, the rich world has committed

Hidden emissions in the global periphery

many shameful acts in its modern history, but how many are genuinely confronted in this way? The logic of Germany's deep engagement with its own legacy is rooted in the idea that by exploring the processes that led to an atrocity, those same pathways might be avoided in the future. To know one's history is to avoid repeating it, but how well do we really know our history, and thus our present?

As I embarked on the train journey back to London that Autumn, I reflected on how we deal with our past. When I was taught about the British Empire in primary school in the early 1990s, I was told simply that having delivered various benefits such as railways to our colonial subjects, we the imperialists decided that the time was right to return the colonies to their people. We have come some way from being comfortable in that kind of assertion, but what endures is the sense of finality, of an international era completed and a retreat into national accountability. This is of course not the way that it is, yet every system set up to monitor environmental performance is designed as if it were. By framing the global issue of carbon emissions around individual nation states,[52] they in reality protect a governed environment, in which the terms of environmental protection project and entrench historical inequalities.

The domestic basis of carbon accounting and environmental regulation gives countries the opportunity to hide the damage their productive processes engender, not by resolving it, but by moving it across a national border and thus largely out of sight. This is a problem that has long been treated as a side issue, a technicality of accounting, but it is increasingly the main issue. As more and more of the natural environment is extracted, packaged, and transported around the globe, national boundaries no longer mark the edge of our environmental influence,

Greenwashing the global factory

or responsibility. Though we continue to think in national terms, we live in a globalised world and a globalised environment. Consumption in one continent is fuelled by extraction and degradation in another. Economic gains in the Northern hemisphere are built on environmental losses in the South. Environmental security in the rich world is funded by profits from businesses whose global Southern workers face great and growing climatic risk.[53]

It is not hard, given this, to understand why the term 'carbon colonialism' has emerged in recent years. Yet, as intuitive as it is, this is a term with many meanings. It has been used by various authors to refer to low-carbon energy infrastructures such as wind farms in rural areas producing power for urban areas,[54] emissions trading through offsetting or carbon credits,[55] carbon capture and storage as a normalisation of extraction,[56] the overemphasis of slash-and-burn deforestation in the Amazon,[57] and the 'outsourcing' of carbon emissions into supply chains.[58] Along similar lines, the word 'colonialism' has also been applied to pollution,[59] a phenomenon increasingly noted in recent years as international waste shipments have increased substantially, almost doubling in the EU since 2001.[60]

Amidst these myriad perspectives though, carbon colonialism (and its associated offshoots) boils down to a single point: frameworks that legitimate the exploitation of one environment for the benefit of another are colonial. On a practical level, this is about hiding: the concealment of environmental impacts due to fossil fuels and industrial processes by moving these impacts from one balance sheet to another, or simply by allowing them to fall between the cracks of environmental accountancy. More fundamentally though, it is about legitimation. Rather than simply logistical cover, carbon colonialism means shrouding

Hidden emissions in the global periphery

extractive processes in the 'moral cover'[61] of market-based environmental logics that anticipate and operationalise colonial entitlements.[62]

As Max Liboiron[63] outlines so well in the case of plastic pollution, technical environmental frameworks legitimate and conceal ecological destruction by setting 'safe' levels that place environmental harms beyond the reach of environmental obligation and concern. The 'assimilative capacity' limit,[64] below which water is deemed able to self-purify becomes a hard ethical marker. Plastic pollution below this limit is effectively 'free', even when you can pull it from the gills of a fish – and it is this same logic that shapes the emission of carbon in the global economy. Simply put, there is carbon that matters, because it is produced domestically, and carbon that doesn't, because it isn't.

Reflecting this, recent years have seen calls for a shift in how carbon emissions are accounted for, from a production-based metric in which only emissions produced within a country's borders are counted, to a consumption-based metric in which emissions associated with imported goods also figure in the total. This, argue its advocates, is necessary to close the carbon loophole in carbon policy',[65] wherein wealthy countries claim successes in cutting emissions, despite increasing the total emissions with which they are associated. However, as the cases above exemplify, the fundamental logic of nationalism undermines their effectiveness in practice.

In the case of garments, this is because of the disconnect between analysis of environmental impacts and *responsibility* for those impacts. It is perfectly possible to trace a supply chain across the globe, even if this may in practice be more difficult than expected, but allocating responsibility is a different matter.

Greenwashing the global factory

Responsibility exists only so far as it is supported by the capacity to enforce it, and carbon-accounting mechanisms invariably prioritise domestic emissions because of the greater capacity to compel regulated parties to comply. Metaphorically speaking, there is no inspector whose jurisdiction extends across the length of an international supply chain. National inspections, added together, should in theory fulfil the same role, but in practice the differences in monitoring capacity along the chain mean that they add up to less than the sum of their parts. Companies know this, governments know this, but international supply chain obscurity is deemed an *acceptable* knowledge gap. This is carbon colonialism.

Similar patterns can be seen, more starkly still, in the case of brick imports. This is, to put it bluntly, a trade that would never be allowed domestically. The lugging of millions of tons of low-value matter across distances so vast that the carbon cost of transporting them exceeds the several days of belching smoke and ash produced in manufacturing them would rarely be viewed as a reasonable environmental cost. The six hundred tons of carbon emitted by each container on its journey from South Asia to Europe means the weight of three and a half houses, or five blue whales' worth *of gas* added onto global heating. Yet since those vast volumes of diesel fumes occur beyond the borders of their destination, they are an *acceptable* carbon cost. If they are produced amidst a density of kilns so great that it contributes to a 'great Asian brown cloud' composed of sooty black carbon that hangs above the continent,[66] then this too is acceptable. This too is carbon colonialism.

At base, then, carbon colonialism is not about flows of goods, which have always happened. It is not about flows of waste,

Hidden emissions in the global periphery

which have always happened too, albeit not on the current scale. It is about how systems set up to protect the environment also demarcate which parts of the environment matter and which do not. Fixing this means recognising and confronting the international, national, and human pathologies that have undermined most efforts up to now.[67] Above all, it means recognising that we are not individual inhabitants of 'our common home', but actors on a stage of deeply embedded power structures and economic flows. There is no one-small-change because each change merely legitimates the structure as it stands. We cannot consume our way out of carbon colonialism.

Five hundred years into a system of global economic domination created according to the terms and priorities of the West, there is no quick fix that can change this, but a better approach is possible. National carbon accounting, especially domestically produced emissions, is an increasingly outdated concept, yet it remains the default position in climate policy discourse. An approach that recognises the environmental cost of the vast economic tendrils sprouting from major economies is fundamental to any serious decarbonisation policy. Yet, despite these data being readily available, climate policy continues to be formulated on nationalist lines. In an era of global climate breakdown, this is as avoidable as it is pointless, yet the persistence of this line of thinking speaks to a centuries-old mindset. In a globalised system of unequal power, it is sufficient simply to outsource environmental problems like carbon, like soot, like plastic. Bring in what is necessary and out, across the border goes (or stays) the rest. This, above all, is carbon colonialism, but it doesn't have to be this way. We already have the ways and means to decolonise the way we measure, mitigate, and adapt to climate change.

Greenwashing the global factory

What specifically, though, might this mean? After all, decolonisation has been a widely used word in recent times. It has been applied to education,[68] politics,[69] and international law[70] amongst many other topics. It has seen land returned to native peoples, artefacts returned, statues of colonisers torn down. Yet decolonisation has not only been about confronting the past, but also interrogating the present: recognising how little has changed from the days of colonial hegemony; how many of the old systems are still in use. Building on these struggles, decolonising climate change means decolonising the imagination[71] in order to decolonise the future. It means reshaping our vision of what a successful response to climate change would be. Not as clean, 'green', wealthy nations consuming goods produced in distant zones of discrete responsibility, but as a community of places, intermeshed by globalisation.

This is a task as sizeable as it is vital, but at its core are three priorities. First, carbon emissions targets based on national production must be abandoned in favour of consumption-based measures which, though readily available, tend to be marginalised for rich nations' political convenience. Secondly, with half of emissions in some wealthy economies now occurring overseas, environmental and emissions regulation must be applied as rigorously to supply chains as they are to domestic production. By adopting these new viewpoints, we can aim towards a final priority: recognising how the global factory manufactures the landscape of disaster. Our globalised economy is built on foundations designed to siphon materials and wealth to the rich world, whilst leaving waste in its place. As outlined in the second part of this book, it now does the same for the impacts of climate change.

Part II

Manufacturing disaster in the global factory

5

Climate precarity: how global inequality shapes environmental vulnerability

Early in the summer of 2014, I was crouching next to a fish stall in the Toul Thom Pong market area of Phnom Penh, speaking with an elderly beggar named Yay Mom. It was early afternoon and the heat was intense. In this, the hottest part of the year, temperatures regularly rise close to 40°C and by May it hasn't rained for 5 months. The soil in rural areas becomes so dry that it begins to rise from the earth and suspended dust becomes a fixture of the atmosphere, endowing the heavy humidity of the Southeast Asian air with still more weight. Heat at this time of year is such a strong presence in everyday life that it takes on its own personality, becoming a protagonist in the stresses and dramas of society. It blocks doors, harries workers, wakes sleepers. A substantial part of life is spent evading its malignant omnipresence. And yet here, in the hottest part of the day, in the hottest month of the year, I had found Yay Mom, slow and frail though she was at eighty-seven years old, making her way around the market in her traditional waistcoat and skirt, krawma scarf wrapped around her head, walking stick and silver begging bowl outstretched for the 100 riel (0.25 cent) notes that stallholders and shoppers would periodically deposit.

Manufacturing disaster in the global factory

After offering her a little money, we sought shade under the awning of a nearby fruit stall. The raw fish nearby sweated on its metal tray and the flies circulated with purpose. I asked Yay Mom what had brought her here, to which she proceeded to relate a personal account of Cambodia's astonishing economic development of recent years. She had always been poor and landless, but had made a living as an agricultural wage labourer in Prey Veng province, until the factories began to open and the young people began to leave rural villages in their hundreds of thousands to staff them. This loss of rural labour in the planting and harvesting seasons at first increased the price of rural labour, but before long it became unsustainable. With too few people left to work the farms and the growing availability of microfinance, farmers instead began to invest in new seeds, fertiliser, irrigation, mechanisation to replace the missing labour. Nobody needed Yay Mom anymore and she was by now too old to get another job. Her husband was dead and she had no children, so she began to beg around her own and neighbouring villages, eventually making it to Phnom Penh, where an old woman can make enough to eat and each night rent a space on the floor in a room full of other beggars. She faced this life with resignation, explaining to me that:

> People look down on me, but it's okay. Now I am very lowly. I don't have any children, I don't have anything, I'm very poor. But when I get some money, I will offer some money to the monks and wish that the next life will not be the same as this life.

By the time I met Yay Mom, I had been working as a researcher in various capacities for several years, meeting hundreds of people living difficult and sometimes distressing lives.

Climate precarity

This project, though, had been emotionally one of the more difficult I had conducted. Even by then I was well used to stories of suffering, but in almost every case it was hope and strength that shone through it. Yay Mom and other elderly alms seekers like her, though, were different. Their melancholic reflections were backwards looking, observing without fondness a hard life lived without satisfaction. Despite their passivity, their stories spoke directly, almost accusingly, to the world's injustices, cutting through the flesh of social discourse to reveal an ugly skeleton of unequal privilege. And it was far from only her. I had spoken a few days before to an old woman with a facial deformity who cried as she explained that all she had ever wanted was a family; a few days later a grandmother abandoned by her children at a Phnom Penh hospital, who when unable to pay the bill was released each day to beg for money, before returning every night to repay a microscopic part of her debt. Even amongst many, these are stories that linger.

I don't mention these here as an introduction to further discussion of begging, a topic on which I have written elsewhere at some length, but to exemplify a wider point about work and the environment. I met Yay Mom in the searing heat of the afternoon not by accident, but by design. It is an easy time to spot beggars to speak to because so few other people are out and about. Shopkeepers remain indoors, stallholders sweat in the shade of large umbrellas, Tuk Tuk taxi drivers string up hammocks in their vehicles to recline in. Office workers return home to eat and sleep. Only the poorest of the poor – the beggars, the independent refuse collectors, the street kids – continue their slow roadside progress. And this is no idiosyncratic quirk of Cambodia. Hidden in that list is a clue to one of the most obvious, yet least appreciated,

Manufacturing disaster in the global factory

aspects of climate change: that everybody's experience of the climate is different. The most observant will perhaps already have noticed a second hidden truth: that an individual's experience of climate change has an awful lot to do with their position in society and, on a basic level, the amount of money they possess.

I'll return in more detail to these examples later on in the chapter, but in the meantime, a geographical segue will help flesh out the universality of this point. So, let's depart for a moment the baking heat of Toul Thom Pong market and arrive instead in the slate-grey damp of London, where I, like many London residents, live in a nineteenth-century Victorian house converted into two flats. When the house was built 140 years ago, insulation was not especially effective and little has been done to improve it in the interim. When the boiler is turned off, the temperature drops rapidly, but fortunately I can afford to keep it on. Not everybody is so lucky. Even in a country as rich as the UK, fuel poverty affects a substantial proportion of the population: 13 per cent in England, 12 per cent in Wales, 18 per cent in Northern Ireland, and 25 per cent in Scotland.[1] For those categorised as fuel poor, keeping your home at a reasonable temperature means making compromises elsewhere: eating less or poorer-quality food, not buying new clothes, going without other basic needs.

For those without the luxury of central heating, the experience of winter is visceral. People living in cold houses explain that 'when it's cold, it feels like the whole house has turned into a big block of ice. It feels like someone is rubbing ice on my nose.'[2] They are constantly aware of the cold in the night because it's 'absolutely freezing. It's horrible being woken up

Climate precarity

with the cold because obviously if you're woken up as an adult then you worry about what your kids are going through.' This is a world away from the temperate world that most people in the UK inhabit; even on the coldest days, transitioning from warm house to warm car, to warm office and back means that the better off only experience cold, as it were, at a distance. It is a moment of frost on the skin, numb fingers on a walk back from the shops. The true, numbing, clawing, penetrating face of cold remains unknown. The increased risk of heart attacks and strokes, the respiratory illnesses, the poor diet due to 'heat or eat' choices, the impact on mental health issues, and the worsening of existing conditions[3] are all absent. And yet outside the temperature is the same.

This, then, is a relationship we are familiar with. The intertwinement of wealth and the immediate experience of the environment is a feature of societies the world over. And it has been so for a very long time. After all, humans are naturally tropical creatures. Our body temperature needs to remain in the range of 36.5 to 37.1°C when we are not in motion. Our skin temperature needs to remain between 31°C and 35°C for us to feel comfortable, which means an air temperature of 27°C or a water temperature of 33°C.[4] Without some form of intervention, the latitudes where many humans now live, above 30° North and below 30° South, are lethal to us for much of the year. Our ability to live in most of the USA and all of Europe is therefore inherently artificial. We have had to make our own microclimates to carry with us around the world, whether in the form of clothing, heating, cooling systems, or social practices, all to keep the skin beneath our jumpers, or underneath our parasols, as close as possible to the narrow tropical band in which we can survive.

Signal and noise

This seems simple enough, but the problem is that if we try to extend our thinking beyond heat, things get complicated fast, something I was about to discover at breakneck speed in January 2019. I was at the time in the midst of an hour of being violently jolted up and down by a modified canoe, whilst attempting to traverse Cambodia's Tonle Sap Lake. Essentially a home-made speed boat, the canoe had been innovatively fitted with a Toyota Camry engine, giving it the power necessary to cross the ten kilometres or so of shallow water in a matter of minutes. It was, in many ways, a brilliant bit of low-tech wizardry, albeit not the most enjoyable one to experience.

In planning their undeniably effective creation it was evident that the designer had considered comfort a low priority and the journey veered from the exhilarating to alarming to painful as the boat crashed over the lake's innumerable shallow waves. Yet even amidst the most abrupt digs of metal into thigh it never ceased to be transfixing. The Tonle Sap is one of nature's wonders. It is fed each year by the only river of its size to reverse its flow throughout the year as glacial melts descending from the Tibetan plateau meet the Mekong river. The river then overflows back inland, swelling the Northeastern Cambodian lake to six times its size, before giving way and allowing the mountain rains to drain finally into the South China Sea.

As our floating contraption began to slow, we were still at the centre of the lake, which even in dry season appears as a sea, too broad to spy the banks at either side. Before long, though, the twisted vines, floating bushes, and sunken trees of our destination began to reveal themselves as the bayou-like marshes skirting the edges of the Peam Bang floating

Climate precarity

community. Within moments, we were subsumed within eerie corridors of buzzing greenery, and after a few minutes of drifting, the first wooden houses of one of Cambodia's most isolated communities began to appear: a few dozen floating households separated from the mainland not only by distance, but also by the cost of traversing that distance. My Camry canoe had made the journey, much reduced in dry season, in under an hour, but the cost of fuel alone was over fifty US dollars. The boats that undertake the journey on a regular basis take far longer and in wet season it is practically impossible, making this community extraordinarily dependent on fisheries. Yet this vital industry was dying before the eyes of the community. Villagers report fish numbers declining to historic lows of 10–20 per cent of their previous level, stifling livelihoods with no feasible alternatives to turn to.

I had arrived in this beautiful, troubled hamlet as part of a team collecting data for the Tonle Sap Authority (TSA), the government arm responsible for managing the Tonle Sap Lake, confident in identifying the cause of this decline. After years of witnessing the impact of climate-linked changes in rainfall in rural agriculture, I expected that a similarly clear body of climate-linked impacts would emerge to guide my recommendations. At first, it appeared as if I were witnessing just that. Hot days were more frequent and intense. Fires were occurring with growing regularity in the 'floating forest' trees that, when submerged for part of the year, form a vital haven for spawning fish. Water levels were so shockingly low that even here, in the centre of the lake, you could wade without getting much of your shirt wet. It all seemed initially to point to the influence of the region's rising temperatures, now estimated at over 1°C above 1960 levels,[5] and yet the picture

Manufacturing disaster in the global factory

refused to clarify. Other factors kept intruding, revealing themselves as too important to exclude.

First of all, there were the fires. Climate change undeniably makes these more likely, but spontaneous fires of this kind were not what I was hearing about. Instead, there were accounts of land clearing, by locals and government alike, who wished to use it for other purposes. Many of the communities living on the lake were in the process of being resettled and some were burning trees to clear land submerged in wet season but farmable in dry season. The next problem was that the abolition in the 1990s of the managed 'fishing lot' system that regulated overfishing and specified which species could legally be introduced to the lake had undoubtedly had a huge impact on fish numbers. Every year more and more trawlers now swept the shallow lake clean with nets strung between two boats, leaving nothing but minnows for the local community. This was environmental degradation on a grand scale, but it was not climate change.

And then there was the elephant in the room: the thing that finally collapsed my hopes of identifying a direct relationship between the changing climate of the Mekong region and the degradation of the Tonle Sap: dams. Over half of the water feeding the Tonle Sap Lake is from the Mekong river, yet the upstream portions of this vital waterway have been dammed to such an extent that downstream water flows have already been reduced by 31 per cent.[6] And this was not all. Even beyond these aggregate figures, I learned during my time exploring the lake how influential these structures had become on the water flows of the region. Cambodia alone had constructed two large dams and six irrigation reservoirs, whilst Laos had constructed sixty-three on its portion of the Mekong alone.[7] This level of

Climate precarity

human influence on this vast and wonderful water system was now so great that it is impossible to tell what it will do next, as a senior official for the Ministry of Water Resources and Meteorology had explained to me just a few days before:

> They have many disaster management committees to train people how to read the land, but the flood risk of a river you cannot predict perfectly … because the natural law is now changed. Upstream there are many developments, especially hydropower, so predicting the water levels is not so accurate. Sometimes, when a hydropower dam such as in Laos, for example, is connected, the water levels change and change quickly, dramatically. Before I used to record one hydropower station and they opened it not according to a regular process. Rather, when they see the clouds getting dark, they open the water gates quickly, so the water level changes quickly, making forecasting water levels a little bit difficult for us and making it difficult for us to run modelling to calculate the level of the river.

A little later on, after a lunch of fish and vegetables with the TSA team in one of the village's floating houses, I sat back to consider the situation. The lake was undeniably in trouble, but teasing out the influence of climate change seemed almost impossible. The region's rising temperatures and unpredictable rainfall were so tangled up in the wider tapestry of regional economic development that it seemed impossible to unpick it. The signal was inseparable from the noise. It was all quite unsatisfying, but before leaving for the mainland we had one further thing to see.

In the centre of the marsh, in the centre of the great Tonle Sap Lake, is an area of still water ringed by floating forest. This lake within a lake is its own ecosystem, Boeng Tonle Chmar. It is a bird sanctuary and a legally protected site, but as a haven for edible snails it is also increasingly the source of one of the

109

only remaining supplies of income for the fish-starved residents of Peam Bang. As we entered the lake, a handful of boats were dotted around the calm waters fishing snails from the water, only to be scattered into the trees a few moments later by the arrival of water police. It was a moment that exemplified the sheer range of influences that shape the natural world. On the one hand, the multiple ratcheting stressors of climate change; on the other, the pushes, pulls, and pressures of human development; and on a third, far smaller hand, the systems set up to protect the environment. Real-world ecologies and lived environments exist at the intersection of these three forces. They are no more outside development than they are outside climate change.

For those reading thousands of miles away from the Tonle Sap's lake-in-a-lake, this might seem rather theoretical, but what makes it so important is that it is truly every day to the people involved. Ask a resident of the Tonle Sap whether they are being affected by climate change and they will mostly say yes, they are. Ask what the impacts are and they may say a range of things, but one of the most common is that fish stocks are low. Ask why specifically and you will often get an answer like the following: 'it's too hot, the fish don't like it, the water levels are so low now and the big boats are overfishing'. This is an accurate and perceptive answer, but it is only in part about climate change. So, has our lake dweller answered incorrectly? No. They have simply interpreted the question in the context in which it was asked. The Khmer word for climate: *arkastheat* (អាកាសធាតុ) does not directly translate to the scientific meaning of the term in English, but includes characteristics also of the word environment, *boresathan* (បរិស្ថាន). In addition, for most people who use it, *arkastheat* also has not yet made the scalar

Climate precarity

leap that the word environment has in global Northern public discourse. Just as it used to in Europe, the climate still generally means your immediate climate: weather, air quality, the quality of the rain.

I will say more on this in the next chapter, but suffice now to say that it is by no means a Cambodian, or even a global Southern characteristic. Climate is as much a cultural concept as it is a scientific one,[8] the product of social expectations as well as direct observations. In the US, for example, the phrase 'April showers bring May flowers' is widely used, so people tend to see sudden rainfall in that month as both expected and in some way positive. It speaks perhaps to both the British mindset and weather that the same phrase is widely used with the second part dropped, whilst in Germany the phrase 'Der April macht was er will', or 'April does what he wants', reflects the wilder, continental climate. And this is a two-way street. Cultural expectations are born of what people experience, but they also structure what they expect and what they look for. Fifteen dry days may pass in a British April unremarked, but the first afternoon cloudburst will see the phrase deployed and the norm reaffirmed.

So to a large extent what people say about the climate depends on their cultural norms, upbringing, sayings, and expectations, but there is also a third dimension: why you are talking to them. Since the idea of climate encompasses so many different aspects of the environment, context invariably plays a role in shaping which aspects are discussed. This may sound rather unscientific, but it is in fact equally true at the highest echelons of science. Take, for example, the following definition set out in the influential Intergovernmental Panel on Climate Change (IPCC) 2007 report on impacts, adaptation, and

Manufacturing disaster in the global factory

vulnerability. Although this may seem tightly defined at first glance, in practice it encompasses an infinite variety of different definitions within it, each selected for their particular relevance to the issue at hand:

> Climate in a narrow sense is usually defined as the average weather, or more rigorously, as the statistical description in terms of the mean and variability of relevant quantities over a time period ranging from months to thousands or millions of years. The classical period for averaging these variables is 30 years, as defined by the World Meteorological Organization. The relevant quantities are most often surface variables such as temperature, precipitation and wind. Climate in a wider sense is the state, including a statistical description, of the climate system.[9]

When ordinary people talk about the climate, or changes to that climate, they are effectively doing the same thing that scientists contributing to the IPCC's gold standard reporting are doing: assessing the aspects that they deem relevant, over a timeframe and scale that they deem to be relevant to them. The difference is that, unlike those scientists, they are not defining their terms, checking their data, or recording it in order to make it comparable, so lay climate knowledge remains subjective and a very poor predictor of climate change in a global or regional sense. The signal of long-term climatic change is simply too weak for ordinary people to distinguish from the everyday noise of weather fluctuations.

The lack of fit between what people say and think about the climate and what scientific models say about it has long seen people's own perceptions of the climate dismissed as junk data,[10] but this is starting to change. It is increasingly recognised that, whilst people are generally no good at teasing out one aspect of their environment from another, they are world experts at

Climate precarity

communicating the complex array of issues that shape their environment *in combination*. No predictive model can get close to approximating this complexity and diversity of how climate change impacts individual lives and livelihoods.

What this means in practice is that if you ask two people how the climate is changing, then the answer you get will depend to a great degree on who they are and what they do. If I were to turn around and ask one of the local people of the Tonle Sap, highly dependent on fishing for a living, how the climate has changed in recent years, they are quite likely to say the wind, because wind is crucial to their ability to fish. If I were to ask the same question to a smallholder farmer in the drought-prone East of the country, a mere two hundred kilometres away, I would be less than a tenth as likely to be told the wind had changed and sixty times more likely to hear they had experienced drought. The reason is that people notice the things that affect them, and the more they affect them, the more they notice. Farmers are twice as likely as non-farmers to experience livestock diseases, three times as likely to experience soil problems, and four times as likely to experience insect infestations.[11] To a large extent, what you do is what you see.

As we juddered back towards the mainland, passing school children sailing canoes home from floating schools, the truth of this seemed self-evident. The experiences we share, the conditions in which we live and work, the particular pressures we face in life, all shape how we perceive the environment around us. A hot day or a rainstorm means something very different to somebody spending it on a boat on the open water than it does to somebody spending it in the shade of a floating house, still less somebody in an air-conditioned hotel room a few miles away. It means something different to the old and the young,

Manufacturing disaster in the global factory

the strong and the frail, the dominant and the marginal. Far from being all in it together, we are each in it alone, carrying our own climate with us.

Climate change precarity

Few moments have brought this home to me more clearly than the morning in February 2018, when I sat speaking to a brick worker and his wife in their small metal house, a few metres from the brick kiln in which they worked. Dara and Bopha related a story familiar to many of those working in the industry, of how they had arrived here several years ago in debt bondage and had become so stuck they could foresee no escape, no relief from conditions they were becoming too old and weak to bear. For most of their lives they had been farmers, living in a rural village not far from where they were now. They had survived the horrors of the Khmer Rouge as young children and suffered through the poverty of its aftermath, thanks to the small plot of land they were allocated by the Vietnamese-controlled government in the 1980s. There were good years and bad years, not least the brutal sequence of floods and droughts at the turn of the century which forced them to sell some of their land, but they had always made it through, until things started to change more recently.

The hardest thing had been the departure of wage labourers from the village. Seasonal farm workers like Yay Mom, who we met earlier at Toul Thom Pong market, would accept payment on trust, or in rice, meaning Dara and Bopha could always count on getting their harvest in, but as fewer and fewer people employed them, their numbers had dwindled. When there were no longer any more wage labourers in the village,

Climate precarity

they tried their hand at planting quick-maturing short-grain rice, buying seeds and fertiliser on credit from local shopkeepers, so that they could plant just by scattering seeds on the field. This initially appeared to work well, but yields soon began to decline as the fertilisers sucked nutrients from the soil. More fertilisers proved a simple enough solution, but the bills were going up, the yields steadily shrinking.

When a lengthy drought had hit five years previously, even hired pumps and hundreds of dollars of petrol couldn't save the crop, the well running dry before harvest time. They would soon run out of rice, but there were more immediate concerns. Shopkeepers still had to be repaid for fertiliser, insecticide, petrol. The microfinance lenders they had borrowed the money for the pumps from began visiting every day, until only one option remained. Dara and Bopha sold their land, repaid most of their debts, and approached the owner of this kiln for a little more. He gladly obliged and at long last they were square with the village and the banks, in debt only to the owner of their workplace. After months of feeling besieged by angry creditors, there was a strange relief in this, despite their having to leave their village, and they had begun their stay at the owner's brick kiln with optimism, hoping to be able to repay their loan before long.

It had not worked out that way. Low wages and ill health meant living on the wages they could earn, let alone repaying anything, proved impossible. As Dara complained, 'when we get sick, we don't work and we ask the owner for small sums of money for health services several times, so the debts increased rather than decreased now'. 'The debt increases gradually every year', Bopha interjected, before Dara elaborated: 'last year, I borrowed two hundred dollars from him, and this year,

Manufacturing disaster in the global factory

a hundred dollars. I cannot repay all this, as I work here just to provide a basic subsistence [living]'. With a laugh, Dara reflected, 'we'll work here forever … we'll work here until the boss closes the brick kiln, or until he cuts down the debts or annuls them because we've been working here such a long time. But I don't know whether that will happen!'

As the conversation continued, the aging couple spoke about work and life in the kiln. How they were never forced to work, but had no income if they didn't, how their debts could be passed on to their children if they were unable to repay them. Underpinning it all, though, was a repeated refrain: 'I'm in pain' interjected one worker from behind Dara. 'I feel a constant pain in my bones', Bopha nodded. I asked her if her condition had worsened and she vigorously agreed. 'I always feel so much pain in my knees and waist that I almost cannot walk, but I try to endure this hardship'. 'As for me', the voice came from the worker listening nearby, 'all I feel is a sharp pain in my knees and waist. I get older and I've worked hard for years, so I always feel sick.' By way of demonstration, Dara pressed a finger to his knee. It sank straight into the flesh as though it were a sponge, reaching an inch deep without resistance.

The physicality of work in the brick industry is one of its defining characteristics. It is not only dangerous, but physically corrosive as well, demanding day after day of repetitive exertion: pain on pain of hunger. All of this stands in stark contrast to cold figures and large numbers with which the impacts of climate change tend to be discussed. Climate change in its scientific sense is – and can only ever be – a statistical index, a deviation of averages from one timeframe to the next. So, when we talk about climate change impacts, the results must

Climate precarity

be extrapolated downwards from this index, or they cannot be linked *objectively* to climate change. Put another way, no single event, hardship, or catastrophe can ever be linked definitively to climate change. Just as with asbestos or radiation exposure and cancer, on an individual basis it can only ever be stated that their presence made what happened next more likely. The result is a schism between the disembodied objectivity of the climate itself and the physical, embodied, tangible subjectivity of the suffering it engenders.

As with the people of the Tonle Sap, this means that many of the subtle, complex ways that climate change becomes interwoven with the economics of everyday life and labour get lost in the statistics. Dara and Bopha's story is a classic account of climate change in the global South, but it would not usually be recorded as such. The loss of their land was as much about debt, mechanisation, and development as it was about climate change, but then so are most of the human impacts of global heating. The changes underway in rural areas that saw Dara and Bopha take on loans, and which also left Yay Mom with few options but to leave her home village to seek alms in an unfamiliar city, were not created out of thin air by climate change, but they were intensified and catalysed by it. Environmental pressures hastened mechanisation, they hastened migration to the garment sector and other industries, and for those who remain they continue to act as a push factor and a stressor on livelihoods.

For the vast majority of people, this is what climate change means. It is not experienced as a catastrophic flood, an unending dustbowl drought, or heatwave that leaves bodies in the street. It is experienced as a ratcheting pressure, an ever-stronger push factor, a reduction in bargaining power, a worsening of the terms of work. In Dara, Bopha, and Yay Mom's cases, these pressures were felt through the lens of agriculture. Droughts, unpredictable rains, and flooding contributed to long-term agrarian transitions and provided the final push to send embattled smallholder farmers over the edge into poverty, debt, and finally exploitative work. In the case of the Peam Bang communities, environmental pressures are experienced through reduced and poorer-quality food, diminishing livelihoods, and through criminalisation by proxy. Yet these are far from the only combinations. Far from it. Economy and environment can interact in an almost endless variety of ways.

To exemplify this, let's stride briefly over the three thousand kilometre distance to Bangladesh, passing the lakeside communities of Peam Bang, the Northwestern highlands of Oddar Meanchey, through the waterfalls of Thailand's Khao Yai national park, around the eerie, vacant modernity of Myanmar's capital Nappidaw, and finally to the brick fields of Narsinghdi, near Dhaka. It is an area that announces itself long before its details can be seen, heralded by hundred-foot-tall chimneys that pierce the flat, watery landscape to expel lazily drifting trails of soot and ash. The design of these kilns is a world away from those of Cambodia. In place of the long, low 'boat kilns' worked by Dara and Bopha – so named for their resemblance to an overturned canoe – these 'bull trench' kilns are a breed apart. Built around an elongated oval with a towering chimney at its centre, they cut such an imposing figure across the landscape that their distinctive silhouettes, part clock, part hippodrome, can be counted from space.[12]

Different though they may look, the conditions of work would be all too familiar to a Cambodian brick worker.

Climate precarity

Most workers are debt bonded and child labour is rife. The work is heavy, hot, dusty, and extremely dangerous. Workers fire bricks by stacking them so that they fill the oval base of the kiln, before lighting fires beneath the rick. Sand is poured over the top of the bricks, not only to keep the heat in, but also to allow workers to patrol the stacks on foot, checking the fires are still burning correctly. The lives of these 'firemen' depend entirely on the rick being constructed correctly. Should the bricks collapse beneath them it means instant death amidst temperatures reaching up to fifteen hundred degrees.

Like their Cambodian counterparts, Bangladeshi brick workers are exposed to the climate much more than most. The heat and humidity of Narsinghdi makes for a sweltering working environment, but this direct exposure is only the latest in a long line of climatic pressures that had led them to the kilns. Like Cambodia, Bangladesh is highly exposed to climate change, but its particular combination of agricultural dependence, flood and drought risk, and low-lying delta topography distinguish it from other vulnerable nations, earning it the title 'ground zero for climate change'. For the hundred million people who live in the Ganges-Brahmaputra delta, this means unpredictable rains, salinated fields, and failed harvests.[13] No wonder, in this light, that Bangladesh and Cambodia have vied in recent years for the title of most overindebted country on Earth, as the ever-growing frequency of risk chips away at the margins of rural livelihoods.

We must not, though, make the mistake of dehumanising these processes. People shape their environments just as much as they are shaped by them. One household's crisis is another's opportunity, and the brick industry, as ever, is alive to the

Manufacturing disaster in the global factory

travails of its neighbours. For each of those farmers whose harvests repeatedly failed and whose debts have risen past the point of repayment, there is a brick kiln owner ready to purchase the ailing land as clay. As a farmer and Narsinghdi businessman called Alamgir put it recently, farmers deep in debt agree to sell their soil to brick kiln owners for a few dollars per metre, which helps to clear their debts, but leaves the surrounding area more vulnerable than ever to waterlogging and drought, and 'in extreme danger' when 'water accumulates in the hole, leaving the surrounding lands unstable and leading to their collapse'. A local farmer, Mohammed, put forward a similar view, explaining that 'brick kiln owners extract soil deeply from the land, which brings a high risk of collapse. We have seen a lot of land collapse due to the extraction of soil in the nearby areas.'

Not only is this an issue for farmers in the immediate term, rendering it 'not possible to cultivate the adjacent land', as a second farmer put it, but the firing of the clay presents worse problems still. Local farmers complain that 'due to the heat of the kiln, the plants in the surrounding area don't grow … the smoke from the kiln covers the trees and they don't fruit or flower', so that 'if one farmer sells soil to the kiln, then nearby farmers have to sell'. And this is before the effect on people is considered. Rahman, the owner of a nearby fertiliser business, explained what living in the local area feels like:

> In the morning, it seems as if a gas has formed. You can't look ahead, your eyes burn a lot. The atmosphere is getting heated due to the smoke from the brick kiln, as a result of which people are getting infected with various diseases. Consequently, the average life expectancy is declining day by day. People are

Climate precarity

getting weak at a young age. Children are being affected the most. Again, those who are a little physically weak are also being attacked.

Listening to the stories of brick workers and nearby farmers in Narsinghdi, what comes across above all is how important people are, not only to warming the atmosphere in the first place, but in determining how that warmer atmosphere affects the people around them. This is no revelation, but it bears repeating. The human economy responds in real time to climatic pressures. In Bangladesh, the floods and droughts are the catalysts that make the first sales of land more likely, setting in train a vicious cycle of waterlogging, land collapse, and crop failure that prompts more sales still in the local area. As more and more people give up on their land, the contagion spreads, intensifying the impacts of climate change over a wider area. More people abandoning farming means higher levels of migration to Dhaka and other urban areas, accelerating urbanisation and increasing the demand for bricks.[14] The brick sector continues to grow, swollen by former farmers and fuelled by former farms, and so the cycle continues.

Processes like these are transforming the world under climate change, reshaping landscapes and the terms of work upon them. Yet crucial to understanding them is that they are not fundamentally different from what came before. Climate change is generating more and bigger storms, worse and more frequent floods, longer and deeper droughts, but this is a question of degree not of kind. Events like these do not trample over existing human systems, but work within and around them, so that the human and natural dimensions of a disaster – be it dramatic and sudden, or slower and creeping – are rarely

Manufacturing disaster in the global factory

completely separable. The climate and its pressures are integral parts of the global economy.

And so, across thousands of miles, three countries, and a dozen stories, we come back, in essence, to a broken boiler. Whether shivering in the safety of a London flat, or braving the frontline of the climate crisis in the searing smoke of the Bangladeshi brick belt, the environment we experience depends upon the context in which we experience it. Monsoon rains, even landslides, mean something quite different to someone surrounded by sturdy walls than they do to a person whose ceiling, like many of those who work in the brick industry, is a single sheet of corrugated iron, lifting with each gust of soaking wind. This is precarity in a quite literal sense, but it is the product of economic systems that generate precarity in all its formulations: low wages, lack of regulation, and a paucity of options. For those straining in the heat of the brick kilns, for those nervously watching broiling clouds rise above tall chimneys, and for Yay Mom, carrying her bowl slowly through the glowering afternoon heat of Toul Thom Pong market, the climate has always loomed larger and been crueller in character than it has for other people. No surprise, then, that it is they who are the first to receive its newly strengthened blows.

Nevertheless, despite the universality of this story – one that resonates from the humid tropics to the damp and chilly temperate zones – it is one that much climate discourse remains blind to. Even if we take the responsibility for generating emissions out of the equation, economic inequality is the single biggest determinant of how those emissions ultimately impact the world's populations. The poorer you are, the more vulnerable to climate change you are. If your livelihood is precarious,

Climate precarity

then you are climate precarious: dashed by economic winds against the rocks of climate change. This is grossly unfair, yes, but it should not be depressing. On the contrary. Two centuries of carbon emissions may have unleashed something beyond our control, but the economy remains subject to our will. Environmental justice, in the here and now, is economic justice. A more equal world is the greatest weapon we have in the fight against climate breakdown.

6

Money talks: who gets to speak for the environment and how

Towards the end of October 2008, I awoke to grey skies above Phnom Penh. My neck was cricked against the plane window, my mouth was dry, and my mood bleak. Gingerly, I replayed the events of the previous night. The kind of party that only precedes an indefinite journey, the morning taxi home, the airport, the thirty hour journey, all blended into one, and they had brought me here, to the skies above a city very far from home. My now wife and I had decided to move here to take up work on a research project about migration and natural resource management, alongside a part-time MA. in Development Studies at the Royal University of Phnom Penh. In our idealistic youth we had wanted to learn development not in theory, but in practice; to learn it by experiencing it on the ground. And so, at the age of twenty-three, as the plane taxied to a halt outside Pochentong International Airport, began my life in research.

We emerged from the airport under looming clouds, into a city just beginning to burst from the decaying seams of its pre-war past. Phnom Penh in 2008 was in the process of transitioning from the 'fourth world'[1] economic marginality of the 1990s into the bustling metropolis of fathomless foreign investment

Money talks

and endlessly sprouting glass and steel it is today. The tallest building in the country at the time was the Sorya mall, near the central market: a seven-storey shopping centre topped by a dusty blue glass dome containing a roller rink and bubble tea shop. Almost every other building was two storeys at most, many still made of wood, but change was in the air. In the four years since I had first visited the city, most of the dirt roads had been tarmacked, streetlights had sprung up in places, and the old 1950s Council of Ministers building, designed by the iconic Cambodian architect Vann Molyvann, was now rubble, the plot still awaiting its new-build successor.

The city was undergoing a fast-forward metamorphosis, cacophonously straining concrete muscles ever upwards and outwards. It was a jungle of angle grinders, headlights, and smoke, the product of geopolitical and global economic forces; of foreign investors seeking rapid returns and of low-income governments like Cambodia facilitating them through mass deregulation. It had begun with the economic opening up of the country in the early 1990s following decades of war, and this bubbling metropolis was now routinely held up by development planners as a testament to the magic of the global market. Yet it does not simply happen. These mass movements of matter and money are the aggregation of millions of deals, big and small, of opportunities, but also injustices and infringements.

Every brick at the centre of those gleaming towers was once part of somebody's rice field. How and why they ended up selling their family land may vary, but you can bet that in the majority of cases they didn't want to. However bright the city lights, most households in Cambodia try their utmost to hang on to their rural foothold, because it is where they see their future.[2] That they could not do so is the result of multiple

Manufacturing disaster in the global factory

factors, from the pressures of climate change, to the rising cost of farming, to family illness, to outright deception of brick kiln owners who will literally carve out every plot around yours until you sell. For every proud, thrustingly vertical urban edifice, there is a ragged horizontal shadow carved into the land, from where the piles of topsoil to drive it skywards were cut; memories of abandoned livelihoods refashioned into monuments.

And this is only the bricks. Between every four walls erected in a developing city, there springs the need for a dozen pieces of furniture, decoration, and embellishment. In Cambodia, as in Southeast Asia more generally, this often means rainforest hardwood, a resource for which Cambodia's opening up to the world economy unleashed an apparently insatiable appetite. The houses of wealthy – and in some cases not so wealthy – Cambodians are decked from floor to wall to ceiling in hardwood; large, intricate statues of animals or religious icons are the luxury of choice, but hardwood is an everyday luxury. The chairs, the benches, the beds, the doors: hardwood. In the Angkor Wat temple complex, even the bins are hardwood.

It looks beautiful and those who have been able to celebrate their share of the Kingdom's growing wealth no doubt treasure it, but like the bricks that drive the city skyward these everyday luxuries are haunted by their own environmental cost. As my plane taxied to the Pochentong arrivals terminal in 2008, Cambodia still boasted a huge wealth of primary rainforests. Only ten years later, a quarter of it was gone, swept away alongside 2 million hectares of national tree cover, in a frenzied decade of unregulated land concession and heavy machinery rolling noisily into pristine forest.[3] To visit the Northeastern highland province of Rattanakiri today is to be confronted with

Money talks

a landscape immeasurably changed after thousands of years of sameness. The seemingly endless jungle has given way to wide expanses of deforested farmland, the red laterite soil exposed to the burning sun for the first time in millennia.

This is the hidden cost of development. The raw materials to construct the fresh new face of urban wealth don't emerge from nowhere. They have to be dug, mined, cut, extracted, dredged. More specifically, they have to be dug, mined, cut, extracted, or dredged from *somewhere*: somewhere that used to belong to somebody, that used to have meaning for somebody else. And this question of meaning cuts to the core of the problem in contemporary environmentalism. Protecting the environment is not only a question of looking after as much of the natural world as possible, but on a deeper level, defining what is valuable, in what ways and to whom.[4] You can't protect what you can't define, but the power to define what needs protection is grossly unequal. When a country like Cambodia signs up to a neoliberal development model, as it has, like so many other poor countries, wholeheartedly done in the last three decades, it not only puts its natural assets on the global market but surrenders control of their valuation.

To exemplify this, consider the case of Bhutan, a country which – almost uniquely in the world – has taken an opposite course to Cambodia, rejecting the pursuit of economic growth in favour of a national strategy devoted to the pursuit of 'the Kingdom of Bhutan's overarching goal of Gross National Happiness enshrined in the Constitution through an inclusive and a vibrant democracy'.[5] One of the poorest countries in the world, ranked 178 out of 216 countries and 66 places below Cambodia, Bhutan has nevertheless sought neither large-scale industrial development nor mass tourism. Indeed, anybody

Manufacturing disaster in the global factory

wishing to visit Bhutan must pay some 250 USD per day for the privilege: a sufficient disincentive for most.

Yet despite this, this Himalayan kingdom has made substantial progress in reducing poverty in recent years from 36 per cent to 12 per cent between 2007 and 2017. And it has also been astonishingly successful at protecting its forests, with some 71 per cent of the country's total area under tree cover.[6] This would seem a cause for celebration to most, but not to everybody. To some sections of the sustainability landscape, such figures are viewed as a disappointment, a missed opportunity. Here, for example, is how the World Bank chose to toast this miracle of modern conservation on Twitter in early 2020:

> 71 per cent of #Bhutan's territory is covered in forest, but with a contribution of only about 2 per cent to GDP per year, the forest sector remains underutilized. How can the country sustainably invest in its forests?[7]

Underutilised. In an era in which global forests, the 'lungs of the world' and the planet's primary carbon sinks, are disappearing at a rate of 10 million hectares each year,[8] the prevailing logic of our global economic system still cannot accept a forest simply *being*, as opposed to being used. Bhutan is the only country on Earth with a negative carbon footprint.[9] Its 3.25 million hectares of 'underutilised' forest actively cleanse the planet's air of 6 million tons of carbon each year, four times what the whole country emits, not only for the people of Bhutan, but for everyone and everything.[10] This might, to many people, seem to be rather extensive utilisation, but the World Bank's slant on the mighty achievements of the Bhutanese forests reflects a foundational girder of global capitalism: that something is not used until it is owned, valued, and paid for.

Money talks

Clearly, this is not how most people think about the environment in Bhutan, or anywhere. Yet despite this it remains a mainstay of environmental thinking. The environmentalist and journalist George Monbiot relates a good example of this in a very different context from the Bhutanese forest: Pumlumon in the Cambrian region of West Wales. Not long ago, he explains, a certain insurance company, beset by expensive claims for flooding in Gloucester, developed a novel plan to cut costs. Rather than continuing to pay out on claims like these, they reasoned, it would be less expensive in the long run to adopt a radical solution: purchasing and reforesting the Pumlumon peak in order to slow down the flow of rainwater into the lowlands and thus reduce the risk of flooding in the long term.[11] It was a brilliant idea. Simple, effective, and, above all, green. There was only one problem: the local people hated it. To them, Pumlumon was a site of immense social, cultural, and personal significance. Rewilding it would fundamentally alter a prized local asset, change its character, reduce their access to it. Deciding they would not stand for this, whatever the benefits, they took their complaints to the council, where the plan was blocked and finally dropped.

That there is more to the environment than money might not seem like the greatest revelation to most, but it nevertheless jars alarmingly with the majority of environmentalism. This is because, outside of notable exceptions such as Bhutan, the majority of environmental planning is underpinned at base by free market logic, with one principle in particular holding an outsize influence: the tragedy of the commons, a hugely influential economic idea derived from the American ecologist Garett Hardin's 1968 essay of the same name.[12]

Manufacturing disaster in the global factory

Hardin's essay describes the overgrazing of cattle on common land, narrating an account in which multiple cattle herders have access to a single, commonly owned pasture for grazing. He describes this process in several stages, at the start of which each herder keeps only a small number of cattle on the land. Soon, though, he shows how each herder realises it would be to his benefit to graze a few more, and then many more. Before long the pasture is overrun with cattle trampling grass and competing for fodder. A tipping point has been reached. Eventually the grass is gone, the soil erodes, and the pasture becomes worthless.

In the light of this story, the World Bank's concern over the Bhutanese forest begins to make a lot of sense. Those behind the policy need not bear any malice towards this untouched ecology. Quite the opposite in fact. If you accept the principles set out in the tragedy of the commons, then this huge, commonly owned resource is not only at risk, but *inevitably* doomed without the intervention of the market. It is a perspective that illuminates a great deal about contemporary sustainability, highlighting as it does how the market is viewed not only as a means to profit from nature, but also to protect it from the rigours of human activity. Private ownership from this perspective means responsibility, value, safeguarding. Non-ownership means decay, misuse, and destruction.

The problem, of course, is that centuries of evidence point to an opposite conclusion. Far from being an account of wanton, unthinking destruction, the history of pre-capitalist common ownership often tells a quite different story, one of sustainability over centuries, even millennia. This has not escaped the attention of environmental historians. As one, Susan Cox, argues, for example, 'the traditional commons system is not

Money talks

an example of an inherently flawed land-use policy, as is widely supposed, but of a policy which succeeded admirably in its time'.[13] As another, Derek Wall, puts it, 'Hardin's interpretation is an abstraction that is altogether alien to any iteration of the commons throughout history'.[14] Examples from contexts as diverse as England, India, Mongolia, and the United States highlight how cultural norms of sustainability have long been effective barriers to exploitation of local natural resources. What Hardin did not account for, in other words, was that, as with Pumlumon, societies that are successful in conserving their environments tend to do so for more than financial reasons. The environment has a deep cultural, often spiritual, meaning that transcends its value as a resource.

I, Nature

This was something that was brought home to me repeatedly during my first months living in Phnom Penh, as rural villagers fought and protested the bulldozers sent to cut and haul away the forestry on which their communities depended. In one such case, in the Northern Preah Vihear province, near the border of Thailand, villagers gathered en masse to block the path of diggers, bulldozers, and other heavy machinery. An entire village in one place, with one purpose, is a powerful sight to behold, still more so when backed by all the tools at their disposal. Arraying their three-wheeled tractors in an impromptu battle line, they blocked the pathway of the Chinese company sent to clear land awarded as part of a large-scale land concession, hoisting placards against the company and its workers. They held a ceremony in the village, in which they lit candles

Manufacturing disaster in the global factory

and prayed for the company directors to be struck by lightning, bitten by cobras, and eaten by tigers.[15]

The villagers' action brought the company back to the table to negotiate. And this was far from the only such case of forest spirits being invoked as bulwarks against the destruction of nature. The well-publicised Pheapimex case of 2009, for example, resulted in five hectares of an economic land concession returned to public ownership following the intervention of a forest spirit. Just as the machines were poised to move into the forest, a woman claiming to be possessed by a spirit of the land, or *Neak Ta* in Khmer, stood up to protest what was happening, demanding that the operators stop clearing the trees in front of the pagoda, and shouting 'This is our home, the mountain has been our home since ancient times. If you cut the trees we cannot stay.'[16]

Workmen were so intimidated by this display that the incursion ground to a halt, with company bosses eventually granting the return of five hectares from the 315 hectare land concession back to the local pagoda where the local spirit had made her intervention. It was another rare victory in the longstanding history of land acquisition by industry, but for every such case there are dozens of others where the spirits are said to have abandoned their human neighbours. When the bulldozers finally roll into forest after forest, villagers decry that the spirits that they had served and respected for so long had succumbed to the bribes and greed of modern industry, that they had agreed to relinquish their own land, and with it the villagers.

This is a story told again and again across Cambodia: the idea that the spirits had retreated to the margins of the country, following the wild beasts that used to stalk the plains into the shadows of the remaining forest. And it is more than just

Money talks

a local fable. This animistic belief in the agency of nature was one of the earliest 'discoveries' of Western Anthropology, dating back over 150 years to the 1870s.[17] Yet it is so widely held around the world that more recent anthropologists have argued that the West is in fact the exception, that the *absence* of animist thinking may be the divergent cultural norm. The reality is that most of the global population do not view nature as inanimate. In the eyes of the majority, there are human qualities in the natural environment, just as there are natural environmental qualities in humanity.

Yet widespread though it may be, this kind of thinking has traditionally sat rather uncomfortably alongside environmental governance in the Western mould, which classifies the natural environment as the private property of people. As transnational legal scholar Susana Borrás puts it, 'this is because the idea of rights is built around property rights, which implies a relationship of superiority between humans and non-humans, as well as an appropriation of nature'.[18] In other words, it is the distinction between humans and nature that is the problem in the first place. If you put one entity in the position of hierarchical superiority over another, then the lower of the two will *inevitably* become exploited. This is an opposite view to environmentalism guided by the Tragedy of the Commons and it flies in the face of the kind of market-based policy exemplified by the World Bank's interpretation of Bhutan. It is not, though, as fanciful or as radical as one might think.

On the contrary, environmental law has been steadily moving in this direction since 1972, when the United Nations Declaration on the Human Environment, or Stockholm Declaration, stated that a person has not only the right to 'satisfactory living conditions in an environment whose quality

Manufacturing disaster in the global factory

allows him to live with dignity and welfare' but also the obligation 'to protect and improve the environment for present and future generations'. The Stockholm Declaration was not legally binding, but was absorbed into constitutional law in the form of fundamental rights absorbed by a succession of states. In 1998, 50 states had recognised the right to a clean and healthy environment. By 2016, this number had risen to 193.[19] In 2021, the UN Human Rights Council adopted a resolution recognising the Human Right to a Healthy Environment, formally writing environmental protection into the foundations of international law.

So, the supposedly immutable legal barrier between humans and their environment is breaking down in one direction, but this is not the only change. In recent years, various legal cases have begun to recognise not only the human right *to* the environment, but the rights *of* the environment itself. One of the first such examples took place, perhaps surprisingly, in California in 1972, in a case called Sierra Club v. Morton, where a company had purchased an old-growth forest in order to turn it into an amusement park. The judge argued that numerous non-human entities, from ships to corporations, assumed the rights of personhood for legal purposes and that nature should be no different. The amusement park was shelved.

And this is not all. In recent years, whole countries have begun to recognise the personhood of the environment. Both Ecuador and Bolivia, in 2008 and 2010 respectively, integrated nature into their constitutions, not as a resource to be protected as property, nor as an environment necessary for the health of humans, but as a subject in itself, with a set of inalienable rights similar to those possessed by a person. In Ecuador, the person even has a name: *Pachamama*, the Quechua equivalent of

Money talks

Mother Earth. She has the right to maintain the balance of her ecosystems and all of her natural cycles, not only to exist, but to regenerate and evolve.

Legal frameworks like this represent not so much a paradigm shift in the relationship between humans and nature as a recognition that a failure to reach the right answers may mean that the questions themselves are the problem. Yet recognising this is not enough on its own to change the terms of the debate. In a globalised world, the questions we ask about the environment are neither locally nor freely chosen. The terms of engagement with nature are set elsewhere and access to the environmental conversation often tightly constrained by economic circumstances. Long before the environment can be spoken for, the question of who gets to speak has already been decided.

Who gets to speak?

A few weeks after arriving in Phnom Penh, as I sat in the morning sun on a bench beside Wat Phnom, the central temple of Phnom Penh, I was blissfully unaware of the unequal politics of environmental knowledge. At this time of the morning it was not yet too hot and the palm trees dotted around the complex still provided ample shade. Not far away, a wizened Asian elephant named Sambo, supposedly a survivor of the Khmer Rouge and still bearing a deep scar on her leg to prove it, waited with her owner for the first tourists of the day to take a ride around the temple on her back. It was the first day of fieldwork for the new project I would be part of, exploring how migrant workers respond to changing environmental conditions. The birds chirped, the motorbikes buzzed around the surrounding road,

Manufacturing disaster in the global factory

and at 9am we two British researchers met our Cambodian counterparts and set off for our first assignment.

Our colleagues, who had already been working on the project for a couple of months previously, started us off easy, talking to a group of motorcycle taxi drivers waiting for customers in the shade around the temple. Leaning on the battered red Suzuki motor scooters that were the ubiquitous urban transport in 2000s Phnom Penh, they were open and friendly, explaining their patterns of migration, their reasons for leaving their rural villages, and their plans for the future. This was research. We were doing it. We were getting data. And through the foggy lens of inexperience, we felt as though we were the first to do it, breaking new ground in scholarship, asking questions that had never been asked before.

And so things continued for two years. Working alongside a local undergraduate researcher, it was my job to work through questionnaires, clarify any ambiguities, record answers, and – once the fieldwork was eventually complete – compile the results into something useful. Other than having an undergraduate degree from the UK and all of two months of a new Master's course under my belt, it was not a job I was especially well placed to do. I did not at the time speak much Khmer and had no fieldwork experience. I had spent a total of six months in the country by this point, but still knew little in depth about it. I was, to put it simply, wholly underqualified to do research: a shortcoming I might have felt more guilt about had not the salary largely reflected my talents: 280 USD per month, including expenses, a sum on which I lived for six months until it was eventually doubled to just over 560 USD for the remainder of my stay.

I detail this because it reflects something fundamental about academic research and indeed about the world of work more

Money talks

generally: gaining experience comes with a huge opportunity cost, which not everybody is willing or able to bear. Those two years of research were not intended as an internship. Indeed, I initially had no special aspiration towards research until I came to love it in the course of doing it. It was simply a way to earn a little money, which was better than no money. At the time it certainly did not feel like a privilege, but, as I would come to realise years later, not everybody has the luxury of low-paid work.

This is a problem brought to prominence by the rise of unpaid internships in countries like the US and UK, which have placed an immovable financial barrier between people from less well-off backgrounds and many of the most desirable jobs. On a personal level, though, this lesson was conveyed to me early on in my time as a researcher, as I reached the end of my first two-year research project. I had worked every day of that period with a Cambodian colleague, Peakhadey, who was only a year younger than me. He was coming to the end of his bachelor's degree and I was beginning a Master's. We were – and remain – good friends and he is every inch the researcher that I am. Yet after the conclusion of that project our lives moved in different directions. I returned to London to begin a PhD, whilst he, still at the time hoping to have a career in research, was sucked into his struggling family business. That lost decade of looking after first his wider family's affairs and then his own family's affairs meant, as for so many people, that the chance of further study was inaccessible, the imperative to provide too pressing to permit the indulgence.

This is how voices are silenced. For Peakhadey, the responsibility to provide for loved ones was too great to take

Manufacturing disaster in the global factory

decisions that leave him poorer in the short term, but this is by no means a problem for Cambodians alone. It is a decision faced by millions of people around the world who are similarly constrained by economic obligations, and it is a major problem for social mobility in wealthy countries. In the UK, for example, almost half of graduates under twenty-four had completed an internship in 2018, of which 70 per cent were unpaid.[20] No surprise, then, that middle-class graduates are 50 per cent more likely than their working-class peers to take them up,[21] leaving a gaping hole in sectors, like education, research, and policy advocacy, that speak for the environment.

Even for those lucky enough to be paid throughout their academic journey, degrees incur debt and salaries are low even for those who secure funding for their doctoral studies or postdoctoral positions. Prospective scientists or environmental scholars have to put up with financial constraints for decades until they reach more senior roles. The result is structural exclusion of the less well off from knowledge production. In the UK, academia is one of the most inequal professions, with over 50 per cent of the sector coming from the most privileged 'professional and managerial' backgrounds, compared with only 15 per cent from working-class backgrounds.[22] The US reports a very similar story, with parental education being one of the most reliable predictors of an academic position. Academic faculty are up to twenty-five times more likely to have a parent with a PhD, a rate that doubles at the most prestigious universities.[23]

And even this is only part of a much bigger story. The world is full of universities, full of students, and full of academics: over 31,000 universities and higher education institutions at

Money talks

the latest count,[24] including over 2,500 in China, more than 3,000 in the United States, and a staggering 5,000 in India. Yet amidst this vast tapestry of knowledge, encompassing tens of millions of senior scholars, only a minute fraction – centred overwhelmingly on the global Northern academic power-houses of Western Europe, North America, and Australasia – possess a global voice. The great majority, by contrast, are confined only to a local influence, excluded from accessing the funding necessary to lead environmental research, or to publish in the places where the audience is biggest.

As the sociologist Fran Collyer, who researches academic inequality, summarises, this results in 'significant inequality in citation counts between global North and South, and show[s] that despite the existence of knowledge production in the global South, Southern scholarship is rarely cited by either Northern or Southern scholars'.[25] Simply put, when global Northern scholars write about the environment, even tropical environ-ments far from where they live and work, they predominantly cite other people like them: the vast majority of whom are not only from the global North, but the West. When Southern scholars write about their own environments, they do the same: cite those same Western scholars talking about their countries, their lands, their landscapes. A recent global study of 26 million papers by 4 million authors found that the top 1 per cent of scientific authors accounted for 21 per cent of all citations in 2015.[26] This handful of scholars, the vast majority of whom are from Western Europe and North America, wield excep-tionally powerful voices in the definition of the world and its environments. Climate knowledge, like any form of knowledge, is power.

139

The power to see

Knowledge is power. This is a phrase we are all familiar with, but it can often seem a little abstract. Whose power? The power to do what? How does knowledge shape the environment? All good questions, so let's tie things to another concrete example: the problem of hydropower dams on the Mekong river. One of the world's longest waterways, at 4,350km, the Mekong river flows through six countries: China, Myanmar, Laos, Thailand, Cambodia, and Vietnam. It is one of the world's most biodiverse rivers, one of its richest inland fisheries, and is integral to the livelihoods of over 60 million people.[27] Yet the Mekong is in dire straits. As the Singaporean Yousof Ishak Institute warned in 2021, 'the Mekong River ecosystem is on the verge of irreversible collapse due to the accumulative effects of climate change and increased numbers of upstream dams as well as other humanmade activities such as deforestation, sand mining, extensive irrigation for agriculture and wetland conversion'.[28]

The Mekong, in other words, faces a tangled web of intertwined environmental issues working in combination to undermine the river both as an ecosystem and as an environmental resource. Any one of these issues would be challenging to resolve in a single country, but the combination of multiple problems across multiple nations presents an impenetrable additional barrier to action. In other words, it is extraordinarily easy to identify problems as they exist in any given part of the Mekong. All it takes is a visit, or a conversation with one of the tens of millions of people who depend on it. I was able to speak to fishermen and women who depended on the Mekong a few years ago and they told an unwavering story of environmental

Money talks

breakdown: fish stocks are down 80 per cent at least, maybe more. Water levels are down to record lows (a fact confirmed by monitoring stations). The river is dying and everybody who can do is leaving.[29]

Clearly, this is a problem, but whose problem? A waterway that stretches across six countries belongs to everybody and nobody. Everybody would like it to be a sustainable resource, but every time others exploit it, the incentive to exploit it increases for everybody else. When damming began in China in the early 1990s, it was at first hailed as a victory for clean energy; a move away from the choking coal power plants that had proven so destructive to the local environment and which continued to contribute so relentlessly to global carbon concentration. As dam after dam rose from the upper Mekong though, this narrative soon began to change. As would soon become increasingly clear, the Mekong was being strangled by the infrastructure sprouting up around it.

This seemed at first impossible. Stretching half-way down the Asian continent, the Mekong river is an almost unimaginably powerful force, yet these dams were built to be a match for it. They are monoliths, the scale of which is genuinely difficult to conceive without seeing it with your own eyes. The largest, the Dachaoshan dam in Yunan province, is a gargantuan feat of engineering: a hulking brown-grey concrete obelisk rising 115 metres above the deep greens of the mountainous riverbanks. At its peak, the Dachaoshan can store almost a billion cubic metres of the Mekong's water at a time, roughly equivalent to drinking water for 2 million people for a year. It is a profoundly powerful object, capable of reshaping nature, yet it is only one of eleven major dams currently operational on the Chinese portion of the river, all of which combine to

Manufacturing disaster in the global factory

squeeze away a third of the mighty Mekong's flow by the time it reaches Southeast Asia.[30]

And yet China is far from the only culprit. Despite their own reliance on the river, the countries of the lower Mekong have begun to dam the river with almost equal enthusiasm since the turn of the century. As part of its plan to become 'the battery of Asia', Laos has constructed sixty-three, mostly smaller, dams on its portion of the Mekong alone.[31] Even Cambodia, thought to be the worst affected by the Mekong's dwindling water levels, has got in on the act, constructing two large-scale dams and six irrigation reservoirs of its own. In total, there are now over one hundred hydropower dams in operation on the Mekong, with a further hundred planned or under construction, a record of human intervention that has transformed not only the ecology of the river, but also the very mechanics of its flow. This gigantic force of nature is now shackled by human hands.

This in itself might sound like a problem. And for the Mekong's countless millions of human dependants, not to mention the ecosystem they depend on, it certainly is. The Mekong region has in recent years been beset by water shortages, with five El Niño events in six years between 2014 and 2019 compounded by enormous volumes of water held in dams upstream. The result has been crippling downstream water shortages. The severe droughts that resulted hit people and communities hard, with repeated crop failures caused by low water levels leaving smallholder farmers with almost no income and often saddled by huge debts, and forcing many families off their land. In 2019, water levels in Cambodia got so low that the country's hydropower dams were unable to produce enough electricity to keep the lights on. Literally. Rolling blackouts were the sweltering

Money talks

everyday reality for six months and as much of the countryside nursed hungry bellies, the cities heaved in discomfort.

On one of these long, hot blackout days, I left my apartment in the South of Phnom Penh in search of answers to the issues the country was facing. Traversing the clanking growl of diesel generators, brought in by many urbanites to provide power during the outages, I weaved my way through the afternoon traffic, passing countless families perched outside on the pavement, braving the thick blanket of exhaust fumes to escape the aggression of the heat. After a short while, I pulled to a stop outside the Mekong River Commission (MRC), the body founded in 1995 as a way for the four states of the lower Mekong – Laos, Thailand, Cambodia, and Vietnam – to share data on hydrological management. I paid the driver and crossed the courtyard into the inviting air conditioning of the office. Government buildings, naturally, have their own power supply.

I was greeted by one of the scientists working on river monitoring, Dr Boran, who invited me into his yellow-walled office. His desk, like those of the others in the room, was stacked with papers, books, and computer monitors, and even in the fierce light of the afternoon, blue glass windows and strip lights gave the room a twilight feel. As he began to explain, what made the drought so difficult to deal with was not only its severity, but also its unpredictability. The Commission was set up to measure water levels, but human intervention on the scale that the Mekong had seen in recent decades had greatly reduced the influence of 'natural' environmental factors such as rainfall and temperature, whilst increasing the role of human decisions over dam operation. In effect, the Mekong was no longer operating like a natural waterway at all:

143

Manufacturing disaster in the global factory

They have many disaster management committees to train people how to read the land, but the flood risk of a river you cannot predict perfectly. ... Why? because the natural law is now changed. Upstream there are many developments, especially hydropower, so predicting the water levels is not so accurate. Sometimes, when a hydropower dam such as in Laos, for example, is connected, the water levels change and change quickly, dramatically. Before I used to record one hydropower station and they opened it not according to a regular process. Instead, when they see the clouds getting dark, they open the water gates quickly, so the water level changes quickly, making forecasting water levels a little bit difficult for us and making it difficult for us to run modelling to calculate the level of the river.

With the influence of human activities this great, those in charge of upstream dams have – in combination – the ability to turn the flow of the Mekong river up and down like a tap. It is a startling testament to the power of human development to influence the environment, but also an indicator of how urgently these human factors need to be taken into account. It is as clear a case as any for the need to share local data on industrial activity and environmental indicators beyond national borders. Yet instead it has become a case study in the opposite: how politics and interests get in the way of open data. Despite its ostensible commitment to data sharing in the Lower Mekong River Basin, the Commission is considered to be amongst the weakest of such schemes in the world,[32] and this was borne out by what I was hearing. As Dr Boran continued:

So now, Cambodia, Vietnam, and Laos are the members of the MRC, but sometimes it is a little bit depressing. ... Even though we are members of the MRC and it is supposed to facilitate exchanging information, sometimes we compete in what we

Money talks

use according to the requirements of economic growth of each country, so it can be difficult.

Leaving the Commission later that afternoon, I turned back for a moment as I strolled across the courtyard. The off-white 1990s building was a government building like any other at the time: the red roof tiles, the deserted foyer, the dusty corridors, the tinted windows. And this was the point: environmental data, the fundamentals of what we know about the environment, are governed by the same rules as everything else. Arrangements for data sharing are intertwined not only in national political structures and international political arrangements, but also wider political discussions over resource use, situating them in the same political negotiations as other issues related to the management of the Mekong, including hydropower, shared economic development strategies, and water management.

This was not only about the Mekong. It pointed to something bigger. The questions we ask about all aspects of the environment reflect the geography of our needs. We may be curious to know the air pressure of a particular cubic metre over the mid-Atlantic to three decimal places, but it is unlikely that this information will ever exist. In general, we know what we need to know. This, in itself, is logical. We can't possibly know everything after all, but it gets more complicated when we start to unpack the 'we' in that statement. Not everybody, not every state, not every region, has the same capacity to create knowledge. The rich world of Western Europe, the US, China, Japan, and others, has a far greater capacity to generate environmental knowledge than do less wealthy countries.

Control over knowledge reflects control over resources. Environmental data reflect power and money, just like

Manufacturing disaster in the global factory

everything else.[33] And this is true of far more than a river. As we have seen throughout this book, almost every country's economy is now so global, so interconnected, that no one country is able to oversee it. Just as it is along the Mekong, knowledge about these trade flows is controlled by different parties, each of whom has their own interests, which shape what they choose to reveal and what they choose not to. Just as no country wishes to reveal that its dams are causing droughts downstream, neither does any country wish to reveal its factories to be flaunting sustainability targets. The incentive for a country like Cambodia, or Vietnam, or even China, to meaningfully 'get tough' on environmental regulation when they are heavily dependent on foreign buyers that advertise a commitment to sustainability, is limited. Buyers, in a similar position, are quite content to leave responsibility with them.

This sheer pace with which human activities are remoulding the natural world is staggering, altering not its form, but its patterns, flows, and habits. This is now a human planet, but perhaps not one as it might have been envisioned when new technology and old power combined in the eighteenth century to set the global economy on its present explosive path. Our human environment is not the product of a man behind the curtain, of the mastery of nature that mid-twentieth-century society once viewed as inevitable. It is the story of innumerable by-products, of uncountable irrelevancies, the boundless legacy of what 'we' don't need to know. The human planet is our unknown planet, the unseen, unintended, and unwanted shadow of everything we do, the Jekyll to the Mr Hyde of economic development.

Conceiving the scale of this environmental spectre can be difficult, so it is something I attempt periodically by taking a

Money talks

trip up one of Phnom Penh's proliferating rooftops. Peering into the illuminated gloom, the city's full panorama of gleaming, empty, skyscrapers is on show, a petrified forest of glass, steel, and blinding light. Somewhere, lost amidst the optical chaos, is a shorter, less imposing, darker building, the Canadia Bank tower with its distinctive metal halo. When this was opened in 2010, a few weeks before I left Cambodia, the bank organised an open party. Not much more than a decade ago, I stood on the roof and looked around at nothingness, marvelling at the unique verticality of my standpoint. In a little more than ten years, a city has been raised from the Earth and the earth, in turn, has lost a city. Just imagine the interior of one of the thousands of gleaming windows that now swamp the humbled rooftop: the glass, the furniture, the smart phones, the electricity that powers them, the water that fills the taps. Where did it all come from? What had to happen for it to arrive here? We can guess, but we'll never know, because 'we' don't need to know.

Even more so than what we do know about the environment, this vast landscape of non-knowledge speaks to the crisis at the root of climate breakdown. It is, at base, not a technical problem, but a problem of power, which makes it a hundred times more difficult to solve, because power is about far more than force. It is about the frames we carry with us to understand the world, what they allow us to see and what remains invisible because we don't possess the tools to know it.[34] Power, from this perspective, looks less like an army and more like a railway track, steaming people along a certain pathway not only because it is habitual, but also because it is the only train we know how to catch. All the vast lands beyond the track, all the people and opinions and facts that make up the world

Manufacturing disaster in the global factory

beyond remain unknown and inaccessible to passengers, who nonetheless feel they are progressing. When it comes to the environment, this infrastructure of knowledge keeps the world moving on its current track, amplifying the voices of the status quo whilst silencing those who think differently. And it is tremendously powerful. The rich world has no need to use force when it retains the capacity to set values. As long as a hectare of land can be sold to a brick kiln for a few hundred dollars, then it will continue to be sold, baked, and elevated into the urban skyline.

Yet there is, as ever, another way. It is possible to reject the globalisation of environmental value, by giving voice to the people it belongs to. Environments do not have to be merely abstract commodities. On the contrary, it takes a lot of work to make them so because this kind of conceptualisation is so foreign to the way that most people think about the environment. Few people look at their garden, their local park, or the hills that surround their village as a potential source of revenue. Their value is evidently holistic and inextricable: you cannot take a part away without ruining the whole. That the same people are happy to use a brick, or a wooden chair speaks to the moral legitimacy that underpins global production. It is assumed that those resources *cannot* have been gathered destructively, that they *cannot* have been treasured in the same way as those with which they can see through their own window. Otherwise, who would permit such expropriation?

The global economy runs on these kinds of assumptions – and in some cases they are true. Not every resource was a treasured personal and community asset. Yet the majority were. And the silence that accompanies their ever-accelerating extraction is not the result of indifference, but of a vast

Money talks

inequality of voice; of people unable to speak for their gardens, their forests, and their hills. This is a central problem in sustainability thinking, which has seen louder and louder calls for a more locally sensitive way of thinking about nature: a 'place-based science' of climate impacts, in which locality and local knowledge carry greater value.[35] Giving greater value to the way that people think about their own local environments is seen as a way to decolonise our environmental thinking,[36] to move away from extractivism and perhaps forestall the slow death of nature that began in the 1700s.[37] Yet, of all the many challenges involved in greening the global economy, this is perhaps the most intractable because it speaks to so many scales of power. How, in practical terms, to bring voice to the voiceless?

This is a thorny problem. Keeping dissenting voices quiet by setting the agenda has always been one of the primary tools of political control over people and environments. And so it remains. As the postcolonial political philosopher Frantz Fanon puts it, 'mastery of language affords remarkable power' because it sets the terms on which a problem is approached, what and who is excluded.[38] Technical frameworks are therefore not only a way of explaining the world, but also of controlling it. Just as you can easily prove the necessity of political domination 'if care is taken to use only a language that is understood by graduates in law and economics',[39] so too is the logic of extractivism made to seem inescapable by the juggernaut of governance that surrounds it.

Yet this, precisely, is what the farmers and factory workers at the coalface of climate breakdown lack: a seat at the table of environmental governance, a means to shape the narrative on climate change, or refute the terms they are given. A handful

Manufacturing disaster in the global factory

of stars, from Vanessa Nakete of Uganda to Nina Gualinga of Ecuador, have helped to bring the issues closest to them to prominence. Yet these are the exceptions that prove the rule. Not everybody, by definition, can make a splash in the media, so hitting back against the slow death of nature requires less exciting, but more powerful, tools to give weight to the words and deeds of the geopolitically marginalised. 'Decolonising the mind', in the words of Kenyan novelist Ngugi wa Thiong'o,[40] is not an individual process. It requires concerted action on multiple fronts, from education, to governance, to culture. Before any of this can occur, though, it means rebalancing the power dynamics between people and corporations by giving marginalised voices the capacity to shape the laws that govern them.

This is a slow and frustrating process, but it is already happening. They may not have the headline-grabbing cachet of the latest green technology, but the new environmental laws, now beginning to emerge both nationally and internationally, are the most powerful weapon in the armoury of environmental defenders, capable of cutting across impossibly unequal environmental odds with the backing of leading nations and institutions. The environmental legislation passed by major economies in the last five years, though weak, imperfect, and partial, has for the first time opened the door for people and communities to hit back against the systems of production that undervalue and degrade their environments. And they are already beginning to have an impact. In 2020, 38 cases were filed against corporations for unlawful environmental abuses worldwide.[41] In 2021, that number was 193, targeted at a broad base of corporations for a range of reasons, including misleading claims over clean

Money talks

energy; proposed investment in carbon-intensive projects; failure to adhere to relevant climate change and environmental regulation; and failure to reduce carbon emissions.[42] Every one of these cases is bringing voice to the voiceless. And this is just the beginning.

7

Wolves in sheep's clothing: how corporate logic co-opts climate action

In October 2021, I exited Glasgow Exhibition Centre station amidst a buzzing, whispering, nervously quiet crowd of journalists, academics, policymakers, NGO staff, and sustainability advocates. Climbing the steps into a crisp autumnal wind, I let myself be carried along by the shuffling, mumbling crowd. Turning into Finnieston Street, we were met with wave after wave of fluorescent yellow jackets, as large a police presence as I had ever seen, marching ostentatiously through the imposing hinterland of brutalist, post-industrial grisaille. With every step the crowd thickened and slowed, and the sense of purpose built. The sound of Ugandan environmental activists performing protest songs against foreign environmental exploitation drifted ever closer on the Western Scottish wind, activists held up signs against the wire grills of the perimeter fence: 'be strong, be brave, be wise'. High above, the vast, staid Clyde docks crane, once built to lift trains destined for the colonies, bore the name of the event the eyes of the world were fixed upon: UN CLIMATE CHANGE CONFERENCE, UK 2021.

Despite having worked in and around environmental research for a number of years, this was my first COP. I had not

How corporate logic co-opts climate action

to a fundamental question: can we continue to increase the amount we produce and consume without doing permanent damage to the planet's ecosystems and those who depend on them?

Those who answer this question in the affirmative may be broadly described as advocates of a green growth solution to climate change. This is a grouping with both a specific meaning, indicating advocates of approaches popularised by Nobel Prize-winning economist William Nordhaus, and in a more general sense advocates of the simple idea that continued economic growth is compatible with long-term global sustainability. This is a vexed but vital question that rests essentially on a single empirical relationship, between economic growth over time and resource extraction over time. When plotted on a single historical graph, these two lines appear almost inseparable. When global growth accelerates, resource extraction accelerates. When growth dips for a shorter or longer period, so too does resource extraction. Over a timeframe of decades or centuries, the case seems unarguable: the more material we extract from the Earth's crust, the faster our economy – measured according to GDP – increases in size.

Viewed this way, there seems to be something of a fundamental contradiction between growth and sustainability, but recent years have seen a powerful challenge to this perspective. Economists associated with the 'green growth' paradigm[1] have argued that decoupling growth and resource use is possible, pointing to evidence from the early 2000s, when the lockstep relationship between economic growth and resource extraction appeared tantalisingly to diverge. This early apparent success added rocket fuel to the concept's uptake, firing it not only to prominence, but also sustained global hegemony. This idea

Manufacturing disaster in the global factory

of sustainably infinite GDP expansion, of economic growth increasingly and permanently 'decoupling' from resource use, is now so powerfully influential that it sits at the root of almost all high-level sustainability policy. More important still, it does so in most cases implicitly, reflected in a shared belief in a future both richer *and* more sustainable.

Not everybody is convinced though. The anthropologist and proponent of degrowth Jason Hickel,[2] for example, argues that the much-lauded decoupling of the 1990s and early 2000s was a short-lived blip, with resource use snapping back into step with growth following the 2008 global financial crash. Indeed, Hickel goes further still, arguing that the widely assumed long-term decoupling of GDP from global resource use will not happen in the future. As he posits, the approximate limit of resource extraction that the Earth's systems can tolerate stands at around 50 billion tonnes each year. Already by 2020 the world was consuming 70 billion tonnes, almost 50 per cent over our global resource 'budget'. At current rates of economic growth, estimates suggest that this will rise to 180 billion tonnes by 2050.[3] Even with maximum resource efficiency, coupled with massive carbon taxes, this could be reduced at best to 95 billion tonnes: still almost double our environmental limits. Viewed this way, green growth is not merely challenging, but physically impossible.

Hickel's solution to this, 'degrowth',[4] wherein resource use *and economic GDP* are scaled back in favour of a redistributive approach to human development, is still relatively novel and not yet a part of mainstream policy thinking. Yet its underlying ethos – the idea that sustainability can only be meaningfully achieved by scaling back consumption – is gaining a substantial popular foothold. Catalysed in particular by the recent rise of

156

How corporate logic co-opts climate action

global sustainability celebrities such as Greta Thunberg, who rails against the 'fairytales of endless growth'[5] that underpin mainstream sustainability thinking, an increasingly influential groundswell of opinion has emerged that the economy may be the problem, rather than the solution to the world's sustainability woes.

Traversing the halls of the Glasgow convention centre, however, this was far from apparent. Despite the portentous atmosphere of change surrounding the conference in the outside world, inside business as usual was everywhere. From the eight-foot gold bear promoting nuclear energy, to the International Maritime Organisation's stand advocating decarbonising container shipping (responsible for 1.9 per cent of all global carbon emissions, more than the entire aviation industry) by cleaning the barnacles from the hulls of its ships. After a while, this starts to feel normal, so it is only upon exiting the rarefied air of the COP 26 blue zone that the sheer radicalism of public feeling on sustainability becomes suddenly and jarringly apparent. The thousands attending a Fridays for Future rally on a weekday afternoon, the tens of thousands of people who participated in the hundreds of fringe events around the city, the hundred thousand who joined the people's march through Glasgow on the middle Sunday of the conference, the millions who rallied as part of the Global Day of Climate Action: these people aren't marching to clean barnacles, they are willing to make sacrifices. Many of them already do. They represent a great many more, but in the rooms where decisions are made: at COP 26, at Davos, at the UN Climate Action Summit, you wouldn't know it, because the debate takes place on different terrain.

Manufacturing disaster in the global factory

Gambling on the rain

Walking home on the first evening of the conference, amidst the crowds jostling through Glasgow, I began to feel a sense of déjà vu. This sense that climate policy is made, perhaps like all high-level policy, in an ivory tower far above – and far removed from – the way ordinary people experience and think about the world, was not entirely new. I had experienced something similar a few years ago, in Northwestern Cambodia. A sudden awareness that the systems we use to analyse and shape the world are far smaller and more subjective than we see them. That the range of options they frame sit like a fragile grid over the broiling complexity with which humans and their environments engage.

My last encounter with this mood had been spring 2019, in the Northwestern Cambodian town of Battambang. Exiting our car after a long journey, I had approached a group of men sitting on *krey* wooden platform beds underneath a shady tree on the outskirts of town. It was a relaxed scene of chatter and slow focus that might be found anywhere in the country, rural or urban, comprising people young or old. Looking a little closer though, this group was a little different. Interspersing the usual array of Chinese smart phones were walkie talkies, notepads, and, as various members of the group showed me, high-tech weather apps showing detailed changes in pressure and windspeed across the region and the local area. There was low talk of wealthy men with large, new, expensive machines, capable of detecting air moisture to the tiniest degree of accuracy. All of this was why I was here, why I had come all this way. This was rain gambling.

As with many of Cambodia's pre-war traditions, the origins of this unusual practice have been lost, but it is generally

How corporate logic co-opts climate action

believed to have originated from Chinese migrants at least a century ago. It is, at base, an almost absurdly simple game: choose one of three daily timeslots, 6am–12pm, 12pm–2pm, or 2pm–6pm. Does it rain or doesn't it? It would be tempting to label it quaint, were it not for the vast sums of money that change hands in the course of a day, often millions of dollars in Battambang alone. It would be tempting to label it charming were it not for the brooding undercurrent of enforcement that accompanies any big, illegal business. In Battambang, in any case, this slow-paced city of crumbling French colonial architecture set along the Sangke river, the luminous green 'ricebowl' of Cambodia, rain gambling is none of these things. It is a way of life, the breadth and depth of its embeddedness in local culture having repeatedly nipped any legal effort to oppose it in the bud.

Despite the contemporary gadgetry employed by today's players, what makes rain gambling special is the endurance of idiosyncratic approaches to rain measurement that can seem almost nonsensical to an outsider. Whether or not it has been deemed to rain is not a question of millilitres in a rain gauge, but of whether sufficient moisture has collected on a stack of thirteen sheets of paper to visibly wet a sheet of blotting paper set out below. This priceless piece of tissue is sequestered in a plastic box atop a jerry-rigged tower, attached to a typical suburban house on the outskirts of town. During rainy season it is accessible only through two locked metal gates and guarded by three men, who are forbidden from coming closer than three metres.[6] The unobservability of this setup, universally agreed but always hidden, demands complete faith be placed in a system that is essentially arbitrary. Questions that might in other weather-related circumstances seem crucial – why this place?

Manufacturing disaster in the global factory

Why thirteen sheets, rather than twelve, or fourteen? – are never asked. The rules simply are as they are, as one betting coordinator explained:

> For the rate, if the sky is good and we bet 8$, they pay us 10$ if we win, meaning if it rains heavily and it is enough water, we win 10$. But if it is not enough water, we lose 8$. They have a measurement glass to determine whether it is enough or not. So even if we know that the rain is going in the right direction, we don't know whether it is enough water or not. If it is not enough, we lose.

And this is only one aspect of the game. Before the rain falls on our thirteen pieces of paper, it first runs down a piece of roofing. In the case of the main stack this is plastic, but other alternatives, such as roof tiles, are available. Roof tiles, especially dirty or old ones, tend to absorb the rain, rather than letting it run, adding a further layer of complexity. How hot is it? How dusty has it been? The science involved in calculating this would be enormously intricate, but it is of course not science. It makes no pretence to the aggregation of knowledge, offers no suggestion of a wider meaning. All the walkie talkies, all the gadgetry, all the weather websites, the neatly inscribed record books, all the inter-provincial collaborations between networks of paid and unpaid cloud watchers; it's just a way to have a little flutter, to pass the time, perhaps make a little money.

Having completed our interviews for the day, I returned with my friend and long-time collaborator Peakhadey to our hotel before meeting for dinner. A little later on, while I waited for him to arrive at a local restaurant, I had time to consider the implications of what we had heard from the rain gamblers. What was so fascinating about the practice was its existence almost as a shadow science, looking and feeling in so many ways

How corporate logic co-opts climate action

like meteorology, or even climatology, but directed towards completely different goals. Sitting there on a pavement table, listening to the clouds begin their first rumbling stirs of the new year, I couldn't shake the feeling that there was something deeper in all this than just a game. That if we make the science we need, then the loudest, but not necessarily the greatest, needs must have a lot to say on what we know about the world.

This seemed all the more important given the way that the 'fight' against climate change had shifted tone in recent years. Since the 2016 votes that brought Brexit and Donald Trump to prominence in 2016, it had become a fashionable political trope not only to question scientific conclusions, but also to rouse a groundswell of scepticism around science itself. This equalisation of knowledge has become a staple strategy of numerous right-wing causes in recent years, from British people having 'had enough of experts'[7] in the run-up to the UK's Brexit election to Donald Trump's assertion that 'the concept of global warming was created by and for the Chinese in order to make U.S. manufacturing non-competitive'.[8]

What makes this trick so effective is that it appeals so well to ideals many people have been starved of in recent years: democracy and equality, a high standing for the ideas of the many. Yet it is, at base, a rhetorical trick and nothing more. Just imagine the effort involved in analysing a single ice core: drilling it from amidst the howling ice of Antarctica and shipping it thousands of miles to a laboratory in which a scientist analyses minute isotopic differences of air bubbles trapped for hundreds of millennia. All of this to achieve one data point in one chart of historic temperature change. The reason that her voice might be louder is that the climate scientist speaks not for herself, but for all of this and more.

Manufacturing disaster in the global factory

The trouble is that this huge and complex apparatus of the global climate compact, 'the vast machine' of environmental science[9] that collects and analyses data, that constantly adjusts, refines, and improves itself, is not in fact a machine. It is a collection of disparate humans and their tools, bound together or separated by geography, institutions, wealth, and politics. Some parts are integrated with all the high technology and rigour one might expect – weather satellites and stations linked directly to meteorological organisations, for example – but others are rougher and readier.

Take rain monitoring, for example. As in many countries around the world, rain monitoring in Cambodia does not resemble the stiffly positioned gaggle of white, high-tech-looking instruments that constitute a modern rainfall station in Europe or the US. Many are still operated by individuals, local people who measure rainfall out by hand and report it. For years, many rain monitors have done so out of a sense of civic duty, setting out at 6am and 6pm every day with a tin can, a small plastic measuring cylinder, and a logbook, all issued by the provincial authorities – and all the same tools that were used to measure rainfall a century earlier. These are no longer the only gauges available of course, but new technology brings new problems. Automatic stations require maintenance and in more remote parts of the country it can be difficult to find the expertise to fix problems when they arise. This means that at any given time a certain proportion are non-functional, creating gaps in the environmental record.

And beyond this, there is a further problem, often overlooked: who pays for this upkeep? In Cambodia, the UN recently funded a renovation of the country's rain-sensing infrastructure, but the government must find the budget to

How corporate logic co-opts climate action

maintain it. As with any budgeting decision, this involves a certain amount of negotiation. Budgets are tight and provincial authorities who manage the top tier of local finance don't want to be overburdened. They also want to be treated fairly, meaning an equal spread of costs between each province, regardless of their geography and size.

The problems this causes had been revealed to me a few weeks previously, when I had paid a visit to the imposing, grey-gated villa which houses the United Nations Development Programme in Cambodia, to speak with a technical expert working on these issues. As he had explained at the time, 'we've figured out that to cover Cambodia completely, we need to get at least two hundred stations and I think so far, we have only a combined hundred stations … and [that] … doesn't necessarily mean that they're functioning'. Standing in the way of better data is an administrative politics of scientific data collection that can be very frustrating for those tasked with collating the data, but at the end of the day he told me, lowering his voice and leaning conspiratorially forward in his chair, 'it goes with the politics, the policies of the ministry'. As long as money is allocated by central government, then so too is decision-making power, an issue which often takes technical and scientific decisions out of the hands of the experts. As he continued:

A working group was set up to discuss precipitation measurement, which depends on agro-meteorological zones. We need at least two to three per province, depending on historical disasters, but population density was not taken into account and different funders fund different parts of the network. So, it was organised by province, but not in the correct way. I kept questioning this decision. Distance and hydrology should figure and hydromorphology is also important.

163

Manufacturing disaster in the global factory

This, too, is science. Though a world away from the space age precision of a meteorological satellite, this messy, political, imperfect world of broken machinery, plastic pots and logbooks, volunteers and missing data is the bread and butter of crucial data collection in many parts of the world. Ugly though it may appear, this, too, is part of the vast machine of climatology, which after all has to get its data from somewhere.

A moment later, Peakhadey's approach jogged me out of my thoughts, turning attention, for the time being, to food. As the evening wore on, though, I couldn't halt the train of thought this roadside reverie had begun. Over dinner, we discussed what we had learned about rain gambling that day. He explained to me about the rain market, how prices rise and fall as darker or lighter clouds change the value of a stake. This was the wisdom of crowds, the magic of the market, the weight of experience, and the power of technology all rolled into one. Above all, it was a world away from tin cans at the bottom of the garden, but one was science, and one wasn't. One mattered and one didn't, but why?

Sometimes on warm evenings, answers don't come as easily as questions, but in the balmy light of the Battambang morning, things seemed clear. Scientific truth does not, as Geographer David Demeritt once put it, emerge like 'a God's eye view from nowhere'.[10] It is drawn from the same messy world of politics and interests and uneven economic power as everything else. The particular rainfall patterns around an unremarkable house on the outskirts of Cambodia's third-biggest city are amongst the most minutely observed and best understood in the world not because they are special, but because they are valuable. Contrast this with the vast swaths of the Cambodian landscape about which very little is known, meteorologically speaking.

How corporate logic co-opts climate action

As an NGO staffer working on flood response had told me just a few weeks previously, even the latest increase in station density still left the network 'fifty kilometres apart, so it might be raining really heavily in between the two stations and you wouldn't know it'.

This matters. Although we've grown used to viewing it at a national, or even a global scale, weather in reality has a far more subtle geography. The models we use to make predictions are based on the coarse resolution of available data: one or two hundred weather stations fifty kilometres apart, for example. In some cases, trend predictions are based on downscaled regional models in which large-scale trends are applied more or less equally to smaller areas. The simple fact, though, is that no organism experiences climate at this scale.[11] The weather as we perceive it, the weather as plants and animals feel it, is determined by a range of small-scale factors from topography to exposure to local tree cover. What's more, these differences are not just subjective, they are measurable. The small-scale topography of hills and valleys influences temperature, and science is only now catching up to the long-held local knowledge that cutting down trees reduces rainfall.[12] Yet this small-scale fine-grained intricacy is largely invisible to climatic modelling, because it doesn't matter, or at least, it doesn't matter enough.

This is no criticism of climatic or environmental modelling, but a more general statement about the nature of knowledge. As scientists, as scholars, as ordinary people, even as infants, we do not collect knowledge for its own sake. We collect it because it is useful, because we need it. This is as true within the plate glass monolith of the World Meteorological Association in Geneva, as it is in the isolated forests of Cambodia's Mondulkiri province – parts of which are more than 150 kilometres from

Manufacturing disaster in the global factory

a rainfall station – or amidst the ultra-dense citizen science of Battambang town. Both people and institutions have limited resources, not only of money, but attention, so both spend their finite resources on what matters. And this is the crux: when we dig down to why one thing or another matters, it is at base never purely scientific. Scientific organisations track rainfall at larger scales because larger scales speak to more of us, but *which* of us also matters. That rainfall station density is ten times lower in the Peruvian Andes than Switzerland (one station per 5,000 square kilometres versus 475 square kilometres) tells its own story: the questions we ask about the environment are the questions posed by money.

The politics of climate truth

This thought had first occurred to me a few months previously, one January afternoon in 2019. I was exiting the National Bank of Cambodia feeling disappointed and somewhat perplexed. I and another colleague had just made a presentation about our Blood Bricks research to bankers, microfinance staff, and economists, and it had not gone down well. Our talk on the relationship between climate change and debt had been politely dismissed as inexperienced and emotional, reliant on individual cases, opinions, and small-scale numbers. In retrospect, this should not have been surprising. After all, the Cambodian government had made similar statements through the press, dismissing the findings of our report. I shouldn't have been surprised either given the audience, who reside in the world of objective numbers and economic facts. To them, a hundred personal testimonies of debt, climate change, and indentured labour was simply a hundred old

How corporate logic co-opts climate action

wives' tales, a hundred opinions. They held no greater meaning than a chat with neighbours, or idle conversations at the school gates.

Two and a half years later, on a much bigger stage and the much colder climate of Western Scotland, I recalled this lesson. The language of finance, of economy, of planning, is not receptive to stories. Not because it is disinterested in them, but because they do not qualify as *truth*. On the contrary, financiers, economists, and planners generally adopt a scientific maxim that explicitly forswears the value of such things. The Latin motto, *nullius in verba*, that bestrides the door of the Royal Society in London, the centre of British scientific excellence, even makes the case directly, roughly translating as 'take nobody's word for it'. It is the cornerstone of the scientific method: that objective evidence and never the nebulous arts of persuasion must form the basis of rational thinking. It is a maxim that has transformed the world, underpinning unimaginable scientific progress since Francis Bacon founded the society in 1660.[13]

Yet the domain of pure reason has its limits. Outside the distilled clarity of the laboratory, scientific thinking in its purest form evaporates. Translating carefully nuanced scientific conclusions into lay and policy language necessitates a goal quite opposite to *nullius in verba*: a persuasive, authoritative argument that will convince an audience to accept an argument without recourse to further evidence. This is a transition that must be managed with great care so that a degree of underlying rigour survives it, yet however close the attention, no approach is without pitfalls. As the IPCC – perhaps the last word in rigorous scientific communication – soon discovered when its reports began to emerge in public journalism, however carefully you

Manufacturing disaster in the global factory

set out your findings, people will interpret them in ways of their own choosing.[14]

In the realm of social science, where there is no laboratory to begin with, the situation becomes messier still. Social science has only recently begun in earnest to get to grips with how climate change is affecting society, but it has already made its presence known. Detailed human studies of how people adapt to the changing climate, how they perceive it, and why they deny that it's changing, have brought a nuanced and much-needed perspective on how climate change manifests in the world. In recent IPCC reports,[15] these social scientific framings have provided crucial evidence around the human impacts of climate change, as well as helping to shape wider strategies around climate change communication. Yet, despite this, the union of the social and the natural scientific is often an uneasy one, involving a wary circling of territory as each side guards its methods and contexts.

This is because social science and natural science are not only different sets of methods, but in many cases are also diametrically opposed ways of seeing the world. This was not always the case but, driven in particular by feminist and global Southern scholars who showed up many of the hard facts of colonial rule for the subjective and power-laden opinions that they were, a new era of social scientific thinking was born in the 1960s which endures to this day. Challenging the all-encompassing objectivity of previous decades, this new era ushered in an ethos based on a key premise: since we, as social scientists, are also people, we cannot truly stand outside the social phenomena we wish to investigate. Though we may seek 'the truth', the questions we ask and the way we interpret what we see are *inevitably* partial. We cannot remove ourselves from the story.

How corporate logic co-opts climate action

So social scientists prefer a little subjectivity, but ultimately if politicians want hard numbers, what's the problem? After all, doesn't quantitative information on as large a scale as possible seem like a good thing? Undoubtedly yes. It is useful, but the trouble is that for many of the people most vulnerable to climate change, personal stories are all there is. Lacking education in many cases and invariably lacking the resources to collect and analyse the kind of large-scale data that would be deemed to be admissible evidence in the realm of high-level climate policy, the people most affected by climate change are effectively locked out of debate over the future of the climate. This is an issue highlighted by Maldives President Mohamed Nasheed's cabinet donning scuba gear to sign an SOS message in front of assembled press. It was an apparently light-hearted stunt underpinning a serious message: though we are in grave danger we are not being listened to, we have no voice.

One of the reasons that President Nasheed's underwater cabinet remains such a memorable image to this day is that it cuts to the core of a wider issue about climate change: our unequal capacity to speak for and about the climate. At COP 26 this was brought home to me in a number of ways, but one of the more direct ones was the absence of Cambodia from the blue zone. Whilst numerous countries were hosting pavilions, Cambodia, as with many other less wealthy countries, had not taken this expensive step towards visibility. The only Cambodians I spoke to during the week were a pair of film makers and environmental activists called Sophal and Somalai, who were screening their film at one of the cultural fringe events that dotted the city throughout the fortnight.

This was the first event that I had attended outside the blue zone and the contrast could not have been starker. Far removed

Manufacturing disaster in the global factory

from the suits and bustle of the conference proper, the screening was taking place in a temporary tent called the Glasgow Landing. It was brutally cold, but with the Covid pandemic still ongoing, the viewing area was permeated with a through draft that chilled the bones. Sophal and Somalai sat shivering in insubstantial puffer jackets at the front of the room, so whilst waiting for their film to begin, I made my way to the side bar for a mug of tea and looked around. Almost everybody was wearing oversized woollen cardigans, and several were shoeless in defiance of the conditions. This was a different world entirely to COP proper, an environmentalism borne from somewhere and something quite distinct. It was rooted in ethics, culture, and above all a rejection of consumption, a rejection in many cases of capitalism itself.

And although this tent may have been, from a COP 26 perspective, on the outskirts of the main debate, it was in reality this ethos that was driving it. None of the hundred thousand marchers through Glasgow the next day would be chanting slogans for green growth. None of them would hoist placards demanding carbon capture. When Extinction Rebellion activists glued themselves to five London bridges in 2018, they did not do so in the name of carbon credits, nor did 4 million global climate strikers take to the streets in support of clean coal in 2019. This visible, powerful minority of environmental activists instead espouse a familiar message: 'Change the system, not the climate'; 'We can't eat money'; 'the government is in bed with climate criminals'; 'we cannot eat coal, we cannot drink oil'. Whatever the specific phrasing, this is a demand to reduce consumption by restructuring the global economy on more sustainable terms, a genuinely radical message, but completely distinct from the blue zone technocracy.

How corporate logic co-opts climate action

After the screening, I approached Sophal and Somalai to congratulate them and let them know I had enjoyed the film. They shivered graciously and we spoke for a while before heading away for some food. Sophal was an environmental activist passionately dedicated to protecting the forest of his home province. In a context in which environmental campaigners are threatened and even killed, as when environmentalist Chut Wutty was shot dead by police in 2012, he had decided that the best way to protect his environment was to document and publicise it. With official channels tightly, even violently, closed, the world of artistic or documentary reportage, amplified by the media for consumption by the global public, is often seen as the only shot at meaningful action. Neither he nor Somalai had ever been outside Cambodia before. They spoke no English, leaving the one British activist who spoke Khmer as their only connection to their surroundings. It was a journey into a cold unknown by two young, brave, principled people, and like President Nashid's media-grabbing underwater cabinet, it was a disruptive strategy: a way to circumvent a blockade of bureaucratic inaction by connecting directly to the outsiders at the core of climate activism.

And ultimately this is the root of the climate conundrum. What makes climate change such a difficult problem to solve is not only the technical difficulty of ameliorating it but also the range of interests involved in persuading coordinated action to go ahead. Many people express support for action on climate change – in the UK, for example, it is the third highest scoring issue in the British Social attitudes survey – but their engagement with the details of what to do about it is often limited. As environmental Geographer Joe Smith and others[16] summarise, the issue is that climate change is 'understood to be urgent

Manufacturing disaster in the global factory

and important, and at the same time is widely seen as boring, difficult and confusing. It poses a global risk, and yet is highly divisive. It represents a defining challenge for our age, and yet it is one that many people choose to ignore and some, even, to deny.' In other words, an issue that tends to be presented as unified, albeit highly complex, in policy circles, manifests in the public sphere as a broad and contested landscape of opinion, filtered through strongly held political beliefs and interests.

The Gateses and the Gretas of this world are not natural allies. The sacrifices they are willing to make and what they hope to gain from climate action are far removed from one another. Their visions of the future are quite different, often mutually exclusive. And herein lies the crux. Much is made of the battle against climate scepticism amongst environmentalists, but whilst doubt and conspiracy undoubtedly rumble on, in reality that battle was won long ago. The climate war, as seminal climatologist Michael Mann[17] terms it, has moved on to new and more subtle territory, sowing discord and disruption from within, rather than engaging in open combat. As with many high-stakes conflicts, it never erupts into open battle, but plays out in the proxy terrain of culture, values, and knowledge.

Since it isn't possible to refute the moral arguments put forward by environmental activists head on – you can't, after all, argue *in favour* of inequality, faster warming, or slower action to prevent environmental breakdown – the debate is moved instead onto the more exclusive terrain of numbers. Yet, as highlighted in the case of Battambang's old rain gamblers, numbers are only part of the story. However high tech the data collection, however impressive the tools, there is a far more subtle landscape of wealth, power, and knowledge that tells us what is important and what is not. In reality the world of

How corporate logic co-opts climate action

climate policy is not a compact, not a genuine consensus, but a battleground and we need to recognise it. Failing to tackle these political dimensions of the climate summarily silences the voices of the poor from the global environmental conversation.

In this, as on so many questions, not everybody gets to speak, but for those who listen there is a simple and effective solution. When you are presented with a narrative on climate change, be it a policy or an argument, or a new framework for understanding the environment, read the interests that underpin it. Think carefully and independently about who will benefit. Familiarise yourself with the arguments and narratives made by big energy companies. They are, after all, written for all to see on the sustainability pages of the websites of Shell, BP, Exxon Mobil, and the rest. So, where you read 'we aim to provide more and cleaner energy solutions in a responsible manner – in a way that balances short- and long-term interests, and that integrates economic, environmental and social considerations',[18] learn to recognise that this means blockage, delay, and filibuster. Where you read 'our purpose – reimagining energy for people and our planet – can help us diversify and decarbonise BP and create real value for our shareholders',[19] see that this means only continued expansion of fossil fuel usage and a lip service investment in green technologies where they happen to be profitable.

This ability to read between the lines of green messaging is especially critical because it gives us, as citizens, the capacity to hold our governments to account. Greenwashing, after all, is not the preserve only of companies, but of a political establishment that has quickly familiarised itself with the tricks of more than half a century of 'sustainable' trade. Even on the grandest stages, the language of the climate emergency, used so liberally by world leaders from Joe Biden to Antonio Guterres, is in most

173

Manufacturing disaster in the global factory

cases a smokescreen; window dressing for the environmental status quo. Once you begin to recognise this, it is difficult to unsee. And few are better practiced with this kind of critical vision than Greta Thunberg, who addressed the gathered crowds on the streets of Glasgow following COP 26's fortnight of high-flown rhetoric, with a statement of characteristic bluntness: 'This is no longer a climate conference', she said of the event the world had been watching. 'This is a Global North greenwash festival.'[20]

Thunberg is far from alone in her cynicism. Green promises without a clear set of rules, mechanisms, and sanctions to deliver them have become a staple of government discourse around the world in recent times. The UK government has been accused of 'greenwashing to the extreme' for giving the go ahead to multiple sites of North Sea oil and gas production under the cover of a sustained rhetoric on the country's green energy transition.[21] Yet it has been nowhere near the most brazen. Brazilian representatives at COP 26 committed to end illegal deforestation by 2028, yet in the absence of an operational plan to deliver on this goal it means nothing. Amazon deforestation had increased by 22 per cent in the year running up to that pledge and 2022 began with the fastest pace of deforestation since 2008.[22] In the words of a Greenpeace spokesperson, 'unlike Bolsonaro, the satellites don't lie'.[23]

We don't always have satellites to confirm our suspicions, but we do have voices to demand proof behind the words. We must be radical, political cynics, possessed of a critical eye and a deep-seated suspicion of green promises. It is not negative, counter-productive, or unkind. On the contrary, reading between the lines of green policy in this way is an essential skill for those who wish to protect the environment, because it is the

How corporate logic co-opts climate action

only way to identify opponents to those goals. The open conflict and doubt sowing of the 1990s has retreated to the margins. Everybody now speaks the language of sustainability. Yet a large and loud proportion of those doing so do not have planetary interests at the forefront of their thinking. The challenge, in the 2020s, is to recognise these wolves in sheep's clothing, to push not just for change, but for the rapid and radical change that is needed as well. Anything less is just gambling on the rain.

8

Six myths that fuel carbon colonialism – and how to think differently

I think the economic logic behind dumping a load of toxic waste in the lowest-wage country is impeccable and we should face up to that. ... I've always thought that countries in Africa are vastly under-polluted; their air quality is probably vastly inefficiently low compared to Los Angeles. ... Just between you and me, shouldn't the World Bank be encouraging more migration of the dirty industries to the Least Developed Countries?

(Lawrence Summers, World Bank Chief Economist, 1991[1])

Pausing for a moment at the top of the dump, I turned to face a digger hauling the latest batch of rejected garments, clothing scraps, string, plastic bags, and hangers up the steep incline of the mound. It looked almost cartoonish in its zippy, unsteady ascent, belying the clanking and grinding of its battered metal body and the puttering black fumes from its vertical exhaust pipe. Dropping its tracks suddenly downwards after a few moments onto a flatter area near the top of the pile, the little digger jerkingly disposed of its pile of detritus, one more indistinguishable pimple on this vast, three-dimensional expanse of rags whose destiny to dazzle for a minute never came.

Six myths that fuel carbon colonialism

Turning away, I was about to head down towards the road when a flurry of movement caught my eye. The dump itself appeared momentarily to stir into life. Faces and limbs began to distinguish themselves, as half a dozen or more waste pickers converged upon the new arrival with burlap sacks. Within moments, the pile had been sifted, sorted, and stowed into one of several bags and the crowd dispersed, dump-soiled clothing merging once again seamlessly into the surrounding waste.

Curiosity piqued, I began to clamber in the direction of the motion. Stepping gingerly across oily mounds of unknown origin and the occasional glint of something sharp, I approached one of the workers, who paused from her collection, a little surprised to see a foreigner walking through the dump under the glare of the midday sun. Setting down her sack for a moment, she told me her name was Sopheap and about the difficulties of working where she did. 'You can make money when it is sunny', she explained, 'but it's difficult when it's raining. You can't make anything – and worst of all is my health. There is so much bacteria around here and I have to go to the hospital constantly. I have no energy and I'm always sick.'

High above Sopheap, the sun was reaching its scorching apogee, rousing the large, overfed flies from their lazy meanderings into a more determined buzzing. The air quivered and bubbled and the scattered cows that roamed our mountain of consumption lowed. Curiosity gave way for a moment to physicality. My head swam, my skin dripped, the bad air felt thick, almost liquid, polluted and sour. I had been here less than half an hour, lifting no more than a notebook, but the sheer oppressiveness of this working environment was uncomfortably apparent. Suddenly the obvious question sprang to mind:

Manufacturing disaster in the global factory

'Why not work somewhere else?' Phnom Penh, after all, is a big city and there are many things to do.

Sopheap's answer belied my own condition. She hadn't fallen on the hardest of times as such. She hadn't suffered the family illness, or the devastating drought or flood that triggers a good many of her Cambodian peers to endure rigours such as these. On the contrary, she was here to build a house, or rather, because she had borrowed money to build a house, not far from here. This kind of work was just enough to pay the microfinance loan that the family had taken on and she could at least live in her half-constructed house for free in the meantime. Her endurance was a means to an end: aspiration rather than desperation, a difficult step on the long road to a better life.

I can almost feel the reader relaxing through the page; and myself, instinctively, too. Well, that's OK then! Many people, after all, endure challenging strictures in the service of a greater good. To suffer for future benefit is not only tolerable, but righteous, perhaps even *good*? Delayed gratification is referred to by social psychologists as the 'master virtue', the 'moral muscle' from which all other good treats begin.[2] Most major religions teach abstinence, self-control, and endurance of suffering as a route to future happiness, and moral narratives of this kind have come to dominate economic and environmental planning.

Herein, though, lies the issue. Sopheap in many ways stands, clutching her burlap sack, at the nexus of our global environmental crisis. She is the citizen of a poor country hard-hit by climate change. She has few assets and high debts. Her health is poor and yet she spends the day exposed to the elements in an unhealthy, even dangerous, worksite. She makes a living from the scraps of an industry that, by any reckoning, is one of the

Six myths that fuel carbon colonialism

most polluting and carbon intensive in the world. In a wider sense, she ekes out a life on the detritus of the rich world's over-consumption, squeezing an existence from the dregs of toxic fashion.

She falls, in other words, within touching distance of a great many of the major sustainability questions of our era. Yet, somehow, she remains outside of every one of them. She lives within the economy of garments meant for export, but informally, living on the supply chain, rather than within it. That same informality denies her an audience to her outdoor labours. Since there is no company at which to point the fingers of exploitation, those endless hours in the oppressive heat are her responsibility alone. As for charity, or its modern incarnation the NGO, it's a non-starter. Sopheap's aspiration denies her that; choices, after all, must have consequences. To paraphrase Helen Fein, she remains beyond our universe of sustainability obligation.[3]

The issue is, though, that most people in the world are in this position. The lens through which we view the impacts of the climate is a narrow one, searching always for those exceptional impacts that are distinct and distinguishable from the business-as-usual of the global economy. As a result, when we are speaking about global climate change, we are, in reality, speaking about only very specific parts of it: the parts we can measure and separate most easily from what we view to be normal. And this, at base, is the core message of this book: that the social narratives we use to interpret, explain, and justify the world play a crucial role in what we do and don't see about environmental breakdown.

Throughout this book, I have detailed some of the many moments, events, and conversations that contributed to

Manufacturing disaster in the global factory

building my own perspective on our globalised environment. It's a viewpoint that took many years of interest and privilege and luck to put together and it is, like any viewpoint, by no means definitive. Yet what is so important about personal experience of global problems is its capacity to challenge grand narratives of environment and development, to identify the many 'black swans' – the exceptions that disprove the rule entirely – that pervade systems purported to be sustainable. Above all, it offers the means to give voice to the voiceless, to speak and write with those at the geographic margins,[4] and to bring the words of those crushed beneath the logics of our global system of extraction out into the open: human faces and human words amplified in opposition to the pervasive rationalities of our globalised economy.

The quote which opens this chapter, written by World Bank chief economist Lawrence Summers in 1991, is a perfect example of this. It is unusual for saying the quiet part quite so loud, but the underlying logic is hegemonic. The UK alone exported almost 700,000 tonnes of plastic waste in 2020, almost 2 million tonnes a day and six times more than it did in 2002.[5] The World Bank's maxim of the early 1990s has not only come to pass, but has accelerated with every passing year, even amongst 'world leading' decarbonisers like the UK. And this is only one drop in the global ocean. The UK is behind the US, Japan, and Germany for its total volume of plastic exports, and plastic, of course, is only one of the many forms of pollution kept at arm's-length from the rich world by the ever-roving tendrils of global production.

By taking the reader on a personal tour of some of these tainted landscapes: the dumps, the fishless lakes, the chemical-scorched fields, it has been my aim in the preceding pages not

Six myths that fuel carbon colonialism

only to show that they exist, but also to exemplify the sheer gulf between the rhetoric and reality of sustainable production. There is, simply put, an awful lot in the landscape of sustainability that is misleading, or flat out untrue; comforting myths that hide the dirtiest parts of the global factory from the eyes of the many people who would be horrified to know the truth. These myths are so widespread that they can feel inescapable. They are like mile-high walls around genuine change and meaningful action. Yet it doesn't have to be this way. The way we view the world is a political choice and like any political choice it can be unmade, if it can first be identified. First though, we must unsettle our certainties, and reveal the blind spots in our understanding of environmental breakdown and the enormous injustices that lurk within them.

Six environmental myths and how to think differently

A few months after meeting Sopheap, back once again in London, I broke from the crowds of the high street to buy some socks from a high street brand. Entering the shop, I was struck by the effort that had been made to ensure a feel of sustainability about the place. Previously employing a typically bold, eye-catching high street palate, the large, strip-lit room was now liberally adorned with muted earth-tone greens, ochres, and browns. The signage, previously designed to 'pop', now had a distinctly faded, recycled feel to it. Blond models lined the walls, frolicking in meadows, wearing bohemian dresses, turning their faces to the warmth of the sun. My local sock shop had, in other words, fully embraced the aesthetics of sustainability.

Manufacturing disaster in the global factory

Wandering amongst the racks, I was reminded of Sopheap and her burlap sack. The idea of environmental consciousness, of what it means to lead a responsible and sustainable life, is constantly shaped and reshaped by the influences around us. Commercial greenwashing was mainstream half a century ago, but over the years it has grown and evolved, becoming central not only to the choices we make on the high street, but also to the limits within which we see these choices. More than simply a sign to indicate sustainability, contemporary greenwash has *become* sustainability for many people, encompassing the full range of possibilities between environmental breakdown and salvation.

The consultancy firm Deloitte,[6] for example, found recently that 85 per cent of UK consumers had made a change to their purchasing habits, from cutting back on single-use plastic, to avoiding brands not seen as sustainable. These findings are broadly mirrored across Europe, where studies[7] suggest that almost two-thirds of Europeans were already taking action on climate change, usually in the form of recycling or reducing the amount of packaging they consume. These are the actions of people who care about climate breakdown and are willing to do what they can to prevent it, but what, if anything, do actions like these have to do with the worst effects of climate change? Almost every day the citizens of Europe are bombarded with images of deforestation, desertification, record-breaking heatwaves, and melting glaciers. Survey after survey indicates high levels of concern over these changes, climate anxiety, climate depression, even climate grief. Yet the response is the same as it ever was: a little change to the shopping basket here or there. If it were not so normal, if the idea that small changes in consumption lead to big changes in society were not so hardwired into public consciousness, the scale of the response would be laughable.

And then it struck me. These two worlds are so far apart as to be effectively irreconcilable. You can't get, conceptually speaking, from the sock rack in a North London clothes shop to an indebted waste picker weaving through the cows to reach a little digger full of scraps. There is no direct line of sight because there is no direct responsibility – and this, in no small part, is the intention. Brands, governments, companies, politicians, NGOs, each of the many organisations with which we share our lives has their own narrative on sustainability because they have their own interests. Wherever we live in the world, each of us is in thrall to these accounts, that define what is right and what is wrong for the environment, and what is possible or impossible to fix it. In what remains of the book I am going to try to take this forward, to be a little more specific about the myths that shape our understanding of the global environment, not only for completeness, but also as a platform for action. After all, we cannot fight what we cannot see.

Myth 1. Climate change is causing more natural disasters

Climate change and increasingly extreme weather events, have caused a surge in natural disasters over the past 50 years, disproportionately impacting poorer countries.

(United Nations News, September 2021)

One of the most widely shared myths in all of climate change discourse is the idea that climate change is increasing the likelihood of natural disasters, a burden that is 'disproportionately'

Manufacturing disaster in the global factory

falling upon poorer countries. It is a maxim uttered by everybody from presidents to professors. Yet it is fundamentally flawed. Climate change is not causing more natural disasters because disasters are not natural in the first place. They do not result from storms, floods, or droughts alone, but when those dangerous hazards meet vulnerability and economic inequality. A hurricane, after all, means something completely different to the populations of Singapore and East Timor. In one case dramatic rain down a windowpane and a few hours spent inside. In the other, life-threatening danger.

This difference, clearly, is no accident of geography, but of a global economy that ensures that some parts of the world remain more vulnerable to climate change than others. The global factory is now so central to the lives of so many that very few livelihoods and very few environments are 'virgin territory'. They are instead remade, degraded, or protected according to the whims of globalised production. Natural disasters are therefore economic disasters: the result both of centuries of unequal trade and of the specific, everyday impacts of contemporary commerce. The raw truth is that even amidst the uncertainty of our changing climate, allowing disasters to happen is a choice that we as a society have made and continue to make.[8]

To illustrate this, let's begin with the humble teabag. It is small, light, and unassuming. It is natural, in the sense that it contains a plant, albeit a processed one. It is also green. In total, it takes approximately 32 kg CO2e to make 1 kg of tea, including all processes related to packaging and transportation.[9] Yet, because we need so little of it to prepare our breakfast brew, it is amongst the least carbon-intensive commodities consumed in the average household. Even if you drink four mugs of tea with

Six myths that fuel carbon colonialism

milk per day for a year, that only adds up to one sixty-mile drive per year in an average car.[10] So, what, then, is the problem? What does a tiny low-carbon teabag have to do with the global climate crisis?

In the case of this particular teabag, an unintended consequence of the global tea trade is increased landslide risk in Sri Lanka. This is because tea plantations, made up of shallow rooted bushes on steep hills and employing a large population of workers, are one of the greatest risk factors for landslides, accounting for more than a third of all such disasters around the world.[11] Since 1990, Sri Lanka has seen its landslide frequency increase by twenty-six-fold,[12] transforming them from a minor disturbance to a major national issue. This is not just the result of bad weather. Eighty per cent of landslides are classed as 'human-induced',[13] with the majority linked to tea plantations. As the climate changes, with rainfall becoming more and more intense in these highland areas, tea plantations like these are a magnet for deadly disasters. They may be green and verdant, but they are as much a product of the global economy as the cities where they are consumed, or the container ships that bring them there.

Yet the next time a deadly landslide occurs you are unlikely to hear anything about the global economy. This is simply not how we see 'natural' disasters. Yet it should be. You only need to look at one of the many maps of future climate change projections, temperature changes, or 'global climate risk' to notice a clear pattern emerging, with the wealthier countries of the Northern hemisphere generally looking relatively secure compared with those closer to the equator. Maps like this are commonplace and familiar. They help us to understand the geography of various pieces of information about the world,

Manufacturing disaster in the global factory

many of them natural – such as rainfall, temperature, or wind speed; and others human in origin, such as wealth, political affiliation, or even press freedom. Yet what maps like these often fail to communicate is the ongoing role of the global factory – and especially the wealthy countries within it – in constantly reinforcing and remaking these vulnerabilities. After five hundred years of global capitalism, the world's landscapes are almost universally in thrall to the demands of markets often very far away.

In a very direct sense, this shows the impact of colonialism on contemporary climate vulnerability. The Sri Lankan highlands would not even grow tea if it were not for the actions of British colonisers in the nineteenth century. With no plantations, landslides would be far less common and the population exposed to them – not needed to pick tea – far lower. Each contemporary landslide therefore has its roots in the lengthy history of domination. Yet there is also a further dimension in play. Sri Lanka has little capacity to set the market value of tea, nor, for the reasons outlined in chapter 2, is it easy to escape from a dependency on primary exports like agriculture and minerals once they become fundamental to an economy. One country alone can't upend the historically entrenched economics of its own environmental vulnerability. And it is perhaps this, above all, that requires recognition in assessing the legacy of colonialism on the environment. Two centuries of carbon emissions may have increased the risk of natural hazards, but five centuries of domination shape the context that they meet.

Six myths that fuel carbon colonialism

Myth 2. We can consume our way out of climate breakdown

We all know plastic straws are polluting our oceans, so the best way to make a difference is by opting for reusable straws instead.

(*Good Housekeeping*, 30 Eco-Friendly Products to Help You Live More Sustainably and Reduce Waste, 2022)

In our globalised world, economy and environment are increasingly inseparable. So, with so much of environmental vulnerability bound up in economic processes and decision making, would this not seem like an argument in favour of the vital importance of sustainable consumption? Well, in one sense, absolutely. The products and services we buy are going to have to be an awful lot more sustainable if we are going to make any headway in reducing the growth of global emissions, let alone beginning to reduce them to net zero. There's a problem, though. Standing in the way of achieving *meaningfully* sustainable consumption is the limited information that we as consumers have about the products that we buy. When we stand in a supermarket aisle, examining an array of packaged products, we have almost nothing to go on, barely any oversight of the impossibly complex tangled mesh of processes that we depend upon to fulfil the basic functions of our lives.

So, let's imagine that you are an environmentally conscious consumer popping into your local supermarket for a few things that you need. You approach the till with three items in your basket: a banana, some batteries, and a packet of chocolate biscuits. Being a responsible consumer, you pause for a moment, taking out your phone to look up the products, but specifics are hard to come by. Looking up your batteries first, which

Manufacturing disaster in the global factory

the label says were made in China, you find numerous reports of exploited factory workers and chemical pollution. You read with some alarm that manufacturing these batteries involves the use of numerous heavy metals including aluminium, cobalt, lead, and nickel, and that the exponential growth of industrial wastewater is toxic to aquatic wildlife.[14] You see the Chinese Ministry of Health announcing that industrial pollution has made air pollution the leading cause of death in the country. Nowhere, though, can you see anything linking these impacts to a particular factory, let alone any information on where this pack of batteries, specifically, was made. Deciding that batteries are clearly too risky, you err on the side of caution and return them to the shelf.

Not to be downhearted you turn to the next product in your basket, your chocolate biscuits, which were packaged in the UK. You're pleased to see that they have also included a second country's name on the packet: Ghana, where the cocoa was farmed. Reinvigorated, you return to your phone, to learn that cocoa production in Ghana is – as elsewhere – associated with environmental pollution associated with heavy metals content in phosphorus fertilisers and the leakage of pesticides. You also learn that for every kilogram of cocoa produced in Ghana, 3.2 kg of carbon are emitted.[15] Peering back at the packet, you see only 100 g of cocoa listed, which gives 300 g of carbon. That doesn't sound too bad, but then the other ingredients are rather heavier. That 250 g of butter looks concerning, and lo and behold that can be up to 15 kg of carbon per kilogram![16] So, from the butter alone that's almost 4 kg of carbon. Before you get to the sugar, you have returned to replace them on the shelf.

Six myths that fuel carbon colonialism

All you have left in your shopping basket now is a banana. Surely a natural and environmentally friendly product if ever you have seen one? You return to your phone one last time to confirm your positive impression. The label says Dole, which has a comforting ring of familiarity to it, but your initial optimism is ill founded. You read in dismay that the banana industry consumes more agrochemicals than any other in the world except cotton, that its thick peel and the market dominance of the vulnerable Cavendish variety means that it requires more insecticide than any other tropical fruit. Worse still, you learn that the use of these chemicals pollutes water supplies, contaminates soils, and can have devastating impacts on the health of workers who earn only 9 per cent of what you pay for a bunch.[17] This was not the information you had hoped for. Defeated, you return your banana to its rack and turn towards the door. You put your phone (16 kg of carbon, toxic cadmium mines) back in the pocket of your jeans (6 kg of carbon, water resource depletion), stepping carefully over a puddle to avoid dirtying your new shoes (15 kg of carbon, methane emissions, and tanning chemicals), and return to your Ford Mondeo (17 tonnes of carbon, lithium poisoning, metal mining, deforestation for rubber plantations) empty handed.

It is, to put it mildly, very difficult to keep track of the tangled web of environmental impacts that our every action generates around the world. This is not to say that we shouldn't try, but rather that we can't possibly do it alone. Ethical marks are available in some cases and vary in their usefulness, but they are not a complete solution, in part because most consumers don't have the time or ability to scrutinise their relative value, but even more importantly because many don't always have the financial resources to pay extra for them. In reality, we would

Manufacturing disaster in the global factory

never spend several hours researching products in a supermarket. Nor would we leave a shop empty handed because of what we found. Pressing everyday needs are simply too great for the huge and complex problem of sustainable production to be solved by consumers alone. Placing the burden of climate breakdown on consumers is as unfair as it is ineffectual. Most ordinary people have not profited from the climate crisis and many around the world are losing heavily as a result. Asking them to spend their way out of a crisis is therefore more of a red herring than a solution.

This was a point made, with some exasperation, by the historian Rutger Bregman at the Davos World Economic Forum in 2019. 'Fifteen hundred private jets have flown in to hear David Attenborough speak about how we're wrecking the planet', he began, before delivering the memorable line: 'but it feels like I'm at a firefighters' conference and no one's allowed to speak about water.' What was the water Bregman alluded to? A topic so taboo that he was not invited back the following year: tax. Tax on the profits of fossil fuels, tax on the highest earners, who contribute the most to carbon emissions, tax on the wealth accrued from both. It is here, not amongst the supermarket shelves, that the money to tackle climate change will be found. Sustainable consumption, for all its positives, is a favoured distraction from those who wish to avoid paying their dues.

Myth 3. Environmentalists are fighting for net zero

MOB MADNESS 'Selfish' eco-mob shuts DOWN Tower Bridge as activist abseils down tourist attraction – causing traffic CHAOS in London.
(*The Sun*, 8 April 2022)

Six myths that fuel carbon colonialism

Freed to cause more mayhem! Police release four of the 15 eco activists they arrested for 40 HOUR oil terminal stunt without charging them as drivers face petrol stations still 'drying up'.

(*Daily Mail* [Australia], 12 April 2022)

Climate zealots versus New York homeowners and workers.

(*New York Daily News*, 2 May 2022)

Consider the above headlines, published within a few weeks of one another in 2022 by major newspapers in the UK, US, and Australia. Now, with those in mind, take a look at the following:

Why net zero is DEAD unless Australians go vegetarian and suffer very high inflation.
[First line: Net zero by 2050? Be prepared for $100 steaks.]

(*Daily Mail* [Australia], 15 May 2022)

Net zero nightmare.
[First line: It has been clear for some time that net zero is as dead as disco.]

(*Spectator Australia*, 14 May 2022)

Net zero is 'dangerous': Job losses from Eraring closure to be higher than forecast.
[First line: Australians aren't so stupid, they know that this greens job revolution promised by the Prime Minster and net zero proponents is not going to happen.]

(*Sky News Australia*, 12 May 2022)

The above headlines are taken from newspapers written in – and about – countries very far away from one another, but those paying close attention may have noticed one or more themes emerging. On the one hand, there is the persistent labelling of environmental protestors as a 'selfish' 'mob' of 'zealots', thoughtlessly impinging on the lives and livelihoods of ordinary people. Then, closely related to this, there is the framing of

Manufacturing disaster in the global factory

net zero pledges as an ideological folly, costly, anti-worker, and 'dangerous'. The similarities between these messages are of course no coincidence. They reflect a widely promoted anti-environmentalism message, favoured in particular in right-wing libertarian policy and press circles, which has been a major feature of Western discourse for decades.

What is more interesting, though, is the way in which this is gone about: apparently blunt, but in reality rather sophisticated and highly effective. We have all heard the phrase 'divide and rule', but its hidden counterpoint – unite and rule – is equally important. After all, it is no good dividing allies, if you don't also present a coherent enemy against which they can be pitched. And this is what is happening right now in the global climate conversation. In the ongoing struggles over environmental policy, 'unite and rule' plays a key role in fostering a narrative of 'environmentalists versus the rest'. It determines the shape of what is possible in the minds of the public, allowing favoured, limited forms of action to be foregrounded in place of the radical action that most environmentalists know to be necessary.

The creation of an eco-mob bogeyman within the ranks of environmental advocates is designed to force a backwards policy step. Rather than debating how far regulation of the underlying structures of our economy should go, we are led back onto a battlefield we thought we had left victorious: the 'fight' over net zero. This is known technically as shifting the Overton Window: moving the dial on the public conversation back onto favoured territory. It is the primary weapon of those opposed to environmental regulation precisely because environmental regulation, supported by 78 per cent of Europeans,[18] is so popular. Mark Dolan, presenter for the

Six myths that fuel carbon colonialism

right-wing British television station GB News, provides an excellent example:

> I don't think in 2019 anyone voted for net zero, and I'm certainly not sure they voted for a £20,000 bill for an eco-boiler or 30 to 35 grand for an electric car. It's already an outrage that the British public haven't been consulted about the expensive green agenda, even if it is the right thing to do, which it might be. ... What's worse is the arrogance of eco-zealots. ... There is a certain religious fervour about some members of the green lobby, that we've somehow got to get all of our energy from windmills as of tomorrow morning and segue to a plant-based diet. ... No wonder so many of these eco-protesters don't have a job. With the vegan diet, they've scarcely got the energy to protest.[19]

Beyond their lyrical bombast, the key to statements such as these is in uniting positions that are in reality opposed. Not only are eco-protestors not the same thing as net zero policy, in reality they are more likely to oppose it than support it. For the reasons outlined throughout this book, net zero is seen as a myth in itself, not an outcome to be disruptively sought out. Numerous climate scientists have said as much themselves, calling net zero targets a 'dangerous trap'[20] that slows necessary decarbonisation efforts by assuming future carbon removal technologies will be invented. In this case, though, there is weakness in unity. By conflating the most conservative and the most radical positions in the sphere of environmentalism, the big winner is the status quo.

And who, ultimately, is the beneficiary of this? Naturally, those who commission the headlines. Both *The Sun* and Sky News are run by Rupert Murdoch's News Corp, whilst the *Daily Mail* and *Spectator* are owned by the 4th Viscount Rothermere and Frederick Barclay respectively. All are

Manufacturing disaster in the global factory

billionaires and all are on record for their lengthy history of climate scepticism. Rupert Murdoch has been quoted as stating that 'we should approach climate change with great scepticism'. Frederick Barclay's *Telegraph* hosted the arch climate sceptic James Delingpole, a non-scientist who has described climate science advocacy as possibly the greatest threat Western civilisation has ever known, whilst Rothermere's *Daily Mail* has published headlines along the lines of 'Forget global warming – if NASA scientists are right the Thames will be freezing over again' for many years.[21]

The recent softening of tone in all these outlets is not indicative of the epiphany that many, including Murdoch, would like the public to imagine. It indicates instead the transition to a new phase of the battle against climate breakdown. Rather than outright denial, these and many other outlets are drawing focus into the long grass of endless debate, discussion, and disagreement, in order to evade the radical, status quo disrupting, rewiring of the global economy that is necessary. It is a tried and tested process of conflation, strawman, and filibuster. Nobody voted to pay £20,000 for an 'eco-boiler'? Of course they didn't. Nobody, in fact, is suggesting that they did. Yet filling the airwaves with claims like this kills time. And time is the one thing we don't have.

Myth 4. We need to secure our borders against billions of climate migrants

If you look across sub-Saharan Africa, there are a number of countries there that all have high birth rates, large-scale poverty, are basically Islamic states, and are also suffering from climate change, because it's not just one element. I noticed recently that there's talk of putting together some sort of Western defence programme, because these countries could and are already becoming a

Six myths that fuel carbon colonialism

seedbed for a new centre for Islamic fundamentalism. And so when you say what's coming, what happens next? What happens next is we'll pour lots and lots of money into sending troops into that part of the world and fighting endless wars as we have in the Middle East. But how do we change the debate? It's so compartmentalised. On the one hand it's climate change. On the other, it's the compassion industry and so forth. We're not going to solve these unless we can draw these arguments together.

(The Right Honourable Lord McNally,
EU Home Affairs Sub-Committee meeting on
Climate Migration, 12 March 2020)

Speaking at the EU Home Affairs Sub-Committee meeting on Climate Migration in March 2020, The Right Honourable Lord McNally voiced one of the most widely publicised concerns about climate change. As he summarised, climate change will distress livelihoods in some of the world's most geopolitically unstable regions, unleashing millions of potentially radicalised migrants onto the global stage. It is a threat that occupies much space in the media. Images of untethered hordes released by climate change are a staple of the news agenda.[22] In a wider sense, it is also a key pursuit of environmental scholarship, which has spent decades pursuing the smoking gun of climate migration: a short, sharp, climate shock that affects a large number of people all at once, in a clearly tangible way.

This might be sea-level rise flooding a low-lying coastal area, or it might be drought and high temperatures turning an agricultural area into a dustbowl. In both cases, the assumption is that the people in these places would leave to migrate en masse, becoming a clear case study of climate change reshaping human society. This is important from a scientific perspective, but also from a communications one. Being able to point unequivocally to the world's first climate refugees would be an important new piece

195

Manufacturing disaster in the global factory

of evidence to pressure governments into action on emissions. Yet it has proved continually difficult, the complexity of human lives and motivations an ever-present barrier to clear relationships.

A few years ago, a breakthrough appeared to have occurred. Researchers working in Bangladesh had been working with phone companies to monitor mobility through mobile phone signals. Following a large cyclone which caused flooding along the coast, the researchers detected a mass movement away from the coast. Thousands of signals were moving away from the coast and, crucially, staying away. This, it appeared, was the smoking gun. Climate-linked displacement digitally recorded in front of our eyes. The excited researchers made immediate plans to travel to the sites that the phone signals settled in order to speak in person to these historic migrants. When they arrived, though, they were surprised to find nobody claimed to be displaced. 'We live here', they all said to the baffled researchers. 'We are fishermen, so when we heard that the storm was coming, we travelled to the coast to moor our boats securely, then came home.'[23]

All of which is a way of saying that humans are resourceful in the face of the environment. After all, we have to be. Our bodies are in some ways pathetically fragile. We can survive an internal temperature of 38.2°C for around 80 minutes and 39°C for around 45 minutes before our bodies start to shut down. The external temperature needed to raise our core temperature to this dangerous level is only around 35°C in a very humid environment.[24]

So we can define very precisely what a lethal temperature is, but the problem is that these kinds of temperature occur all the time, even in densely inhabited areas. During recent heatwaves in South Asia, reporters scoured the streets of 49°C Islamabad

Six myths that fuel carbon colonialism

looking for the bodies, but of course there were none. People find shade or a breeze, they drink water, they find fans or air conditioning, so the number that die is far smaller than the huge figures we would expect looking at temperature alone. Life may become worse, harder, poorer, and more unequal, people may get sicker and their bodies weaker, but where there are resources, people will always find a way to adapt.

And the same goes for water levels. We can't live for more than a couple of minutes under water, and not more than a few days even floating on it, but the simple equation that water equals no humans? Not so much. A quick look at the millions of stilt houses around South and Southeast Asia puts paid to that idea. Humans are clever, adaptable, and resilient. They find a way, whenever they have the means to do so. More often than not the human impacts of climate change are defined more by the resources available to adapt to it than the environment itself.

Two people might live in adjacent rooms in the same block, for example, one possessing a floor fan and one not. For these two people, the lived experience of heat is completely different, the impact on the fanless individual – discomfort, dehydration, and sleepless nights – far more severe. And this is no hypothetical. In many hot countries electricity is prohibitively expensive. Many people are unable to afford to use a fan, even if they own one. From the perspective of their bodies, they are living in a different environment from those who can.

The problem, though, is that the way that we talk about the impacts of climate change is not well suited to incorporating this kind of detail. Climate change is a phenomenon born on a spreadsheet, so when it comes to measuring it, everything derives from the large-scale statistics that underpin it. When it comes to the impacts of climate change on *people*, though, this large-scale thinking obscures a great many of the factors that shape experience of the environment. Questions over how the economy shapes these impacts are sidelined, downplaying the importance of economic justice in favour of consumption-led sustainability and techno-fix solutions. When it comes to adapting to the impacts of climate change, there is no solution that comes close to the effectiveness of making poor people less poor.

All of which is why the apocalyptic climate migration narrative is so damaging. On the one hand it plays into harmful political stereotypes around migrants more generally: the familiar narratives and imagery of desperate and dangerous hoards is applied to climate migration just as liberally in the left-wing press as it is to immigration discourse in the right wing of the media.[25] More importantly still, though, it distracts from the action most urgently needed. Rather than high walls to protect against impending catastrophe, it is economic justice that is needed. The 'great displacement'[26] has become such a staple part of climate narratives across the political spectrum that its lack of inevitability has been forgotten. A sea wall can protect against a flood, a fan can protect against a heatwave, a canal can protect against a drought. And money is at the root of all of these. Rather than stockpiling power and resources against a coming onslaught, the rich world can and should be solving these problems through a fairer deal in the global factory.

Myth 5. Sustainability begins at home

You can make these cuts in pollution while driving jobs and growth: we have cut our greenhouse gas emissions by 44 per cent in the last thirty years while expanding our GDP by 78 per cent. And we will now go further by

Six myths that fuel carbon colonialism

implementing one of the biggest nationally determined contributions – the NDC is the pledge we ask every country to make in cutting carbon – going down by 68 per cent by 2030, compared to where we were in 1990.

(Boris Johnson, Plenary Speech to World Leaders at COP 26, 22 September 2021)

Speaking to the United Nations general assembly a few months prior to COP 26 in Glasgow, the British Prime Minister Boris Johnson stood before the assembled delegates and delivered a speech intended to set the agenda on climate change. Amidst the usual classical allusions and highfalutin metaphor, Mr Johnson set out his key aim for this crucial forthcoming conference. Speaking of carbon emissions, he underscored the importance of individual nations setting and achieving their own domestic carbon emissions targets, of agreeing 'to the pledge we ask every country to make in cutting carbon'. It was a speech reflecting a national perspective on carbon accounting that is hardwired into the way we think about emissions. Each nation releases its own statistics and graphs and claims successes or failures based on these figures. And this is not surprising. After all, most of the ways in which we think about global affairs are organised around the idea of the nation state. It is our fundamental political unit, the container within which the rights and obligations that structure our lives play out, whether democratic, or closer to autocracy.

The reality, though, is that economies around the world have outgrown the national systems that govern them. Every country on Earth can either point to substantial domestic emissions reductions (as in the case of the UK or EU, for example), or of being on carefully planned emissions curves, which will achieve net zero by a future date (China and India, amongst many others). This has now been the case for some years, yet throughout this period of great environmental achievement by

Manufacturing disaster in the global factory

major industrial economies, the growth rate of carbon in the atmosphere has increased from 1.5 parts per million per year in the 1990s, to 2 parts per million per year in the 2000s, to 2.5 parts per million per year in the 2010s.[27]

So, if all of the major economies are doing so well at meeting their emissions commitments, why is the atmosphere doing so badly? In a nutshell, because the way we count carbon no longer fits the way we produce goods. With rich countries doing an ever-diminishing share of their own manufacturing, the responsibility to report real-world emissions is left to international corporations with very little incentive to report accurate information on their supply chains. Clearly this is not good enough. In order to meaningfully track carbon, we need to take a global view on supply chain emissions that will make rich countries look far worse than they do at present.

Naturally, this is not a popular point to make to wealthy governments, so when you try to make this case, you will be told, inevitably, that there is only so much a single country can do; that political authority extends only to the borders of one nation and cannot be used to demand change further afield. Yet this, in reality, is a myth. Rich nations have a great deal of control over the goods that cross their borders and can demand any number of conditions to be attached to them. Just think of the stringency with which food imports are regulated when crossing the US, Australian, UK, or EU borders. Any imports must undergo rigorous inspections and the need for independent certification of compliance with a long list of regulations. These standards are in place to protect the domestic environments of these wealthy nations. They are burdensome for companies but they are in place because they are necessary. It is entirely within these nations' gift to apply the same rigour to overseas

Six myths that fuel carbon colonialism

carbon emissions and environmental degradation. That this does not happen tells you everything you need to know about the priorities of rich nations' environmental policies, which are set up to protect home territory, even at the expense of the global environment.

Myth 6. Climate science is an apolitical consensus

The science is finished.

(*An Inconvenient Truth*, 2006)

Climate science, it is often said, has reached all of its necessary conclusions. The Intergovernmental Panel on Climate Change has achieved a consensus, supported by anything from 97 per cent to 99.9 per cent of scientists.[28] So, when Al Gore's global hit documentary *An Inconvenient Truth* proclaimed 'the science is finished' in 2006, it was in a great many ways undoubtedly right. Very, very few scientists would now question whether human-made carbon emissions were warming the atmosphere. Almost as few would question whether we are already seeing the impacts in our seas, our air, our forests, even our workplaces. Yet, despite being right in sentiment, this was the wrong message. The science was not finished, because science is never finished. Questioning, testing, and doubt are part of what it means to be a scientist. Yet this is only half of the problem. Climate science can never be 'finished', not only because it is science, but also because it is politics, as is any endeavour – however objective in its methods – that involves human beings.

Climate change, as a physical process, is a fact: an objective truth discerned by rigorous methods and backed by a huge

Manufacturing disaster in the global factory

body of evidence. Human activities are resulting in ever-larger volumes of carbon emissions into the atmosphere, which is causing global temperatures to rise. This much we know, but scientists don't determine the ways that we respond to climate change, and even if they did, they would be doing politics, not science. Green growth, tougher borders, and military spending in the Middle East are all climate policies, but they are climate policies that benefit particular interest groups at the expense of others.

Understanding how politics and power relations shape media and political thinking is key to achieving meaningful action on climate. The extraordinary achievements of climate scientists in recent decades have brought major strides in understanding the changes wrought on our global environment by human behaviour. It was – and remains – a fight to air an inconvenient truth around the global economy: that the white heat of progress concealed a hidden shadow of destruction. Yet the problem lies in how those very achievements are articulated by politicians and big business, who use the gold standard objectivity of climate science to add weight to policies that protect some people, whilst intensifying the threats faced by others.

So in order to see it differently, in order to imagine different and more effective solutions, we must get political about climate science. Achieving meaningful action on climate change means bringing the voices of scientists, social scientists, environmentalists, and policymakers from outside the rich world into the climate conversation. This means turning to the people currently left to the margins of climate scholarship – the young, the poor, those outside the global North, and to a disproportionate extent, even the female – and hearing what they have to say.

Six myths that fuel carbon colonialism

Exporting the climate crisis

As I descended the side of the dump, the shortcomings of contemporary climate policy seemed as glaringly obvious as they would do months later opposite a North London sock rack. If everything I was seeing, everything I was hearing, was invisible, then the need to change the way we look at our impacts on the world could scarcely be clearer. Mulling this as I trod the disorderly scraps, I was jolted from my reverie by a call from a small brick house I hadn't noticed on the way up, roughly constructed and positioned a few metres from the edge of the high-sided mound. Sitting on a wooden *krey* daybed beside the entrance was a smiling, well-fed man in early middle age, who turned out to be the dump supervisor. Apparently unperturbed by the foreigner emerging from the rubbish, he invited me to sit with him, an offer I gladly accepted after the depleting heat of the exposed dump summit. We sat in the shade of his two-roomed house where he lived with his wife and children, and over the next hour, Norin, as he introduced himself, instructed me in the ways of the dump.

Norin told me how the hundred or so workers collecting from the dump sell what they find to the dump owners to be recycled, but that many workers sneak what can't be recycled off site to sell to brick kilns, or other factories that need cheap and combustible material. It was a little different from the more systematic flow of material I had heard about from other workers, and didn't quite tally with the vast quantities of material that I had witnessed being thrown into burners, bag after bag, or stacked up in high mounds amongst the bricks for later use. Nevertheless, I gave Norin the benefit of the doubt for the time being, and, as I later found out, a large part of the clothing that

Manufacturing disaster in the global factory

ends up in those burners is siphoned off before it reaches the dump itself, so perhaps there was some truth in what he said, or perhaps, after all, there wasn't.

Not long afterwards, I said my goodbyes to Norin, knowing then, as I know now, that I may never understand the full story. Production is quite simply a fiendishly complex business, fading at its edges from the neat, clearly defined formality of the global scale to a far messier, dirtier, and more complex reality on the ground. Recognising this is crucial to tackling the climate crisis, because it illuminates how vital proper regulation is to knowing, let alone controlling or mitigating, the environmental impacts of consumption. It means that we can't get away with just one method to understand the tangled complexity of the economic world we've built to sustain ourselves. We need to confront the vast machine of the global factory on its own terms. That means boots on the ground, it means eyes inside depots and workplaces, and above all it means responsibility. Meaningful action on climate change means closing the complexity loophole: no longer allowing those at the top of the tree of production to thrust their obligations into the long grass of the global economy.

Above all, though, it means seeing the climate crisis through the eyes of those labouring in the global factory: not as a warming atmosphere searing and soaking vulnerable continents, but as a ratcheting pressure on livelihoods already squeezed to breaking point by the pressures of an unequal economy. If there is one message to take from the many stories recounted in this book, it is that the climate never acts alone. When the climate meets humans it does so dressed in the garb of society: economies and systems of governance on the one hand, but also norms, morals, and beliefs on the other. Both these

Six myths that fuel carbon colonialism

hemispheres of the human experience act together to shape who suffers most, who suffers least, and who comes away a winner from climate breakdown.

This is vitally important, not only because it helps us to understand the geography of climate vulnerability, but also because it reveals a new and hugely powerful lever to reshape it. If the changing climate manifests through our social and economic systems, then we can do an awful lot to mitigate its impacts by making changes to those systems. Higher wages means lower vulnerability to climate change, better working conditions means reduced vulnerability to climate change, workers having a voice means far fewer suffering in silence. And all of this stems ultimately from responsible production, changes that begin at home not with individuals, but with laws and regulations that enshrine the duty that companies and the countries they supply have to their supply chains.

This is not easy to do because it is not, at first, easy to see. In order to arrive at the above six myths, the chapters in this book have narrated a personal, fifteen-year journey of learning, seeing, and talking with the people on the frontline of environmental breakdown. Few people have the opportunity to peer over the fence of a garment factory, or to sit and listen to the life stories of those pushed to the limits of endurance by their economic and environmental exposure. Fewer still of those trapped in this vice are able to narrate their experiences, but if they could, if we could hear and we could see, then we would be demanding an end to the outsourcing of climate breakdown.

This is a task with a great many dimensions. Yet at base it is a simple message: people are not vulnerable to the changing climate by accident. They are vulnerable because society makes them vulnerable. Forests, fields, and oceans do not happen to

Manufacturing disaster in the global factory

be vulnerable to pollution and degradation. They are vulnerable because the global factory makes them vulnerable. Not long ago, not in the distant past, but now. The environments of the rich world are – with some bumps along the way – becoming cleaner and safer, even in an increasingly uncertain environment. The resources needed to tackle the challenges of climate change are accruing and being spent to protect their privileged populations. Yet for most of the world, the opposite is true. Natural resources continue to flow ever outward, with only meagre capital returning in compensation. Forests are being degraded by actors big and small, as both climate and market combine to undermine traditional livelihoods. Factory workers are toiling in sweltering conditions. Fishers are facing ever declining livelihoods.

This, then, is carbon colonialism in its fullest sense. Global supply chains are continually reshaping economic conditions, contributing to economic precarity that impedes local efforts to adapt to the changing climate. Workers in the exploited margins of the global factory are facing high levels of exposure to climatic hazards and little capital coming back to help them cope with them. In their local environments, citizens of the global South – those who provide the human and material resources that keep the global factory running – face intensified and complicated risks, shaped by local environmental degradation linked to international production processes. That these supply chains often lead to the rich world presents both responsibility and opportunity: the necessity to act, but also the ability to do so.

This is about more than counting the emissions we use, rather than the ones we produce within our borders. It is about recognising how the economy we depend upon shapes far-off

Six myths that fuel carbon colonialism

environments and above all about taking responsibility for those impacts. As long as it is cheaper and easier to push production beyond the purview of environmental laws, then it will continue to happen. The lack of legal frameworks governing global supply chains means the processes and practices that underpin our lives are in effect owned by no one, responsibility passed up and down the chain ad infinitum. No law can function effectively with such giant loopholes undermining it. So, if there is to be any serious effort to address the impacts of the global economy, then it is this, responsibility, that must be the priority.

We have, in other words, all the tools we need to solve climate breakdown, but lack control or visibility over the production processes that shape it. For decades, this has been frustrating and disabling. Yet change is finally at hand. From legal challenges, to climate strikes, to new constitutions, people are waking up to the myths that shape our thinking on the environment. They are waking up to the 'fairy tales', the greenwashing, and the individualism. They are waking up to the fact that climate change has never been about undeveloped technologies, but always about unequal power. As the impacts of climate breakdown become ever clearer, this has the potential to be a moment of political and social rupture, of the wheels finally beginning to come off the status quo. Yet to succeed it needs everybody behind it, everybody pushing, questioning, and scrutinising. The task is too big, too important, to leave to the goodwill of corporations. So, demand an end to the excuses. Demand an end to the delays. Demand an end to tolerance for the brazenly unknown in our economy. Demand an end to carbon colonialism.

Notes

Chapter 1

1 Althor, G., Watson, J. E., and Fuller, R. A. (2016). Global mismatch between greenhouse gas emissions and the burden of climate change. *Scientific Reports*, 6(1), 1–6.

2 Intergovernmental Panel on Climate Change [IPCC] (2018). Global warming of 1.5°C. *An IPCC Special Report on the Impacts of Global Warming of 1.5°C Above Pre-industrial Levels and Related Global Greenhouse Gas Emission Pathways, in the Context of Strengthening the Global Response to the Threat of Climate Change, Sustainable Development, and Efforts to Eradicate Poverty*. Geneva: IPCC.

3 Hooijer, A., and Vernimmen, R. (2021). Global LiDAR land elevation data reveal greatest sea-level rise vulnerability in the tropics. *Nature Communications*, 12(1), 1–7.

4 Asian Development Bank [ADB] (2013). *Beyond Factory Asia: Fuelling Growth in a Changing World*. Manila: ADB.

5 Thwaites, T. (2011). *The Toaster Project: Or a Heroic Attempt to Build a Simple Electric Appliance from Scratch*. San Francisco: Chronicle Books.

6 Varman-Schneider, B. (2018 [1991]). *Capital Flight in Developing Countries*. London: Routledge.

7 Morgan, K. (2004). The exaggerated death of geography: learning, proximity and territorial innovation systems. *Journal of Economic Geography*, 4(1), 3–21.

Notes

8 Open Apparel Registry (2021). *The Open Apparel Registry Database.* Accessed on 23 December 2021 at [http://openapparel.org].

9 Sainsbury's, J. (2021). *Sustainability.* Accessed on 23 December 2021 at [www.about.sainsburys.co.uk/sustainability].

10 Greenpeace UK and Runnymede Trust (2022). *Confronting Injustice: Racism and the Environmental Emergency.* London: Greenpeace and Runnymede.

11 Hamilton, C. (2006, February). *The Dirty Politics of Climate Change.* Speech to the Climate Change and Business Conference, Hilton Hotel, Adelaide, 20 February 2006.

12 Glazebrook, T. (2008). Myths of climate change: deckchairs and development. In Irwin, R. (ed.). *Climate Change and Philosophy: Transformational Possibilities.* London: Bloomsbury, 162–179.

13 Hickel, J., and Kallis, G. (2020). Is green growth possible? *New Political Economy*, 25(4), 469–486.

14 Greer, J., and Bruno, K. (1996). *Greenwash: The Reality behind Corporate Environmentalism.* Penang, Malaysia: Third World Network.

15 Hickel, J., and Kallis, G. (2020). Is green growth possible? *New Political Economy*, 25(4), 469–486.

Chapter 2

1 United Nations Environment Programme [UNEP] (2020). *Sustainable Trade in Resources: Global Material Flows, Circularity and Trade.* Nairobi: UNEP.

2 World Wildlife Fund [WWF] (2020). *Living Planet Report 2020: Bending the Curve of Biodiversity Loss.* London: WWF.

3 Wilk, R. (2007). The extractive economy: an early phase of the globalization of diet, and its environmental consequences. In Hornborg, A., McNeill, R., and Martinez-Alier, J. (eds). *Rethinking Environmental History: World-System History and Global Environmental Change.* London: Rowman and Littlefield, 177–196.

4 Bales, K. (2016). *Blood and Earth: Modern Slavery, Ecocide, and the Secret to Saving the World.* New York: Random House.

5 Bales, K. (2012). *Disposable People.* Berkeley: University of California Press.

6 Premchander, S., Poddar, S., and Uguccioni, L. (2019). Indebted to work: bondage in brick kilns. In Campbell, G. and Stanziani, A. (eds).

Notes

The Palgrave Handbook of Bondage and Human Rights in Africa and Asia. New York: Palgrave Macmillan, 389–414.

7   Decker Sparks, J. L., and Hasche, L. K. (2019). Complex linkages between forced labor slavery and environmental decline in marine fisheries. *Journal of Human Rights*, 18(2), 230–245.

8   Bunker, S. G. (2007). Natural values and the physical inevitability of uneven development under capitalism. In Hornborg, A., McNeill, R., and Martinez-Alier, J. (eds). *Rethinking Environmental History: World-System History and Global Environmental Change.* London: Rowman and Littlefield, 239–258.

9   Sparks, J. L. D., Boyd, D. S., Jackson, B., Ives, C. D., and Bales, K. (2021). Growing evidence of the interconnections between modern slavery, environmental degradation, and climate change. *One Earth*, 4(2), 181–191.

10 El Kallab, T., and Terra, C. (2020). The colonial exports pattern, institutions and current economic performance. *Journal of Economic Studies*, 48(8), 1591–1623; Bunker, S. G. (2007). Natural values and the physical inevitability of uneven development under capitalism. In Hornborg, A., McNeill, R., and Martinez-Alier, J. (eds). *Rethinking Environmental History: World-System History and Global Environmental Change.* London: Rowman and Littlefield, 239–258.

11 Hickel, J., Dorninger, C., Wieland, H., and Suwandi, I. (2022). Imperialist appropriation in the world economy: drain from the global South through unequal exchange, 1990–2015. *Global Environmental Change*, 73, 102467.

12 Bunker, S. G. (2007). Natural values and the physical inevitability of uneven development under capitalism. In Hornborg, A., McNeill, R., and Martinez-Alier, J. (eds). *Rethinking Environmental History: World-System History and Global Environmental Change.* London: Rowman and Littlefield, 240–241.

13 Mekong Fish Network (2020). *Trash Under the Surface of the Tonle Sap River.* Accessed on 8 February 2022 at [http://mekongfishnetwork.org].

14 Felipe, J., Abdon, A., and Kumar, U. (2012). *Tracking the Middle-Income Trap: What Is It, Who Is In It, and Why?* Levy Economics Institute, Working Paper Number 715.

15 Tilley, L. (2021). Extractive investibility in historical colonial perspective: the emerging market and its antecedents in Indonesia. *Review of International Political Economy*, 28(5), 1099–1118.

Notes

16 Dosi, G. (2016). Beyond the 'magic of the market': the slow return of industrial policy (but not yet in Italy). *Economia e Politica Industriale*, 43(3), 261–264.

17 Doch, S., Diepart, J. C., and Heng, C. (2015). A multi-scale flood vulnerability assessment of agricultural production in the context of environmental change: the case of the Sangkae River watershed, Battambang province. In Diepart, J.-C. (ed.). *Learning for Resilience: Insights from Cambodia's Rural Communities*. Phnom Penh: The Learning Institute, 19–49.

18 Short, P. (2005). *Pol Pot: Anatomy of a Nightmare*. New York: Macmillan.

19 World Bank (2014). *Clear Skies: Cambodia Economic Update*. Phnom Penh: World Bank.

20 International Labour Organisation (ILO) (2019). *Cambodia Garment and Footwear Sector Bulletin*, Issue 9, July.

21 World Bank (2022). *World Bank Databank*. Accessed on 18 May 2022 at [https://data.worldbank.org/country/KH].

22 Human Rights Watch (2015). *'Work Faster or Get Out!' Labour Rights Abuses in Cambodia's Garment Industry*. USA: Human Rights Watch.

23 Derks, A. (2008). *Khmer Women on the Move: Exploring Work and Life in Urban Cambodia*. Honolulu: University of Hawaii Press.

24 Asian Centre for Development [ACD] (2020). *A Survey Report on the Garment Workers of Bangladesh*. Dhaka: ACD.

25 Lee, E. S. (1966). A theory of migration. *Demography*, 3(1), 47–57.

26 Blake, W. (1808). And did those feet in ancient time. In Blake, W. *Milton: A Poem in Two Books*. London: Associated University Presses.

27 Beckert, S. (2014). *Empire of Cotton: A Global History*. New York: Penguin Random House USA.

28 Ibid.

29 Ibid.

30 Ibid.

31 Wittering, S. (2013). *Ecology and Enclosure: The Effect of Enclosure on Society, Farming and the Environment in South Cambridgeshire, 1798–1850*. Macclesfield: Windgather Press.

32 Dwyer, M. B. (2015). The formalization fix? Land titling, land concessions and the politics of spatial transparency in Cambodia. *Journal of Peasant Studies*, 42(5), 903–928.

33 Diepart, J.-C. (2016). *They Will Need Land! The Current Land Tenure Situation and Future Land Allocation Needs of Smallholder Farmers in Cambodia*.

Notes

MRLG Thematic Study Series #1. Vientiane: Mekong Region Land Governance.

34 Merchant, C. (1980). *The Death of Nature: Women, Ecology, and the Scientific Revolution*. London: Harper Collins.

35 Hickel, J., and Kallis, G. (2020). Is green growth possible? New Political Economy, 25(4), 469–486.

36 Szolucha, A. (2018). *Energy, Resource Extraction and Society*. London: Routledge.

37 United Nations Environment Programme [UNEP] (2016). *Global Material Flows and Resource Productivity. Assessment Report for the UNEP International Resource Panel*. Nairobi: United Nations Environment Programme.

38 Global Waste Cleaning Network (14 November 2021). *Global Waste Trade and its Effects on Landfills in Developing Countries*. By Mariam George.

39 Cotta, B. (2020). What goes around, comes around? Access and allocation problems in global North–South waste trade. *International Environmental Agreements: Politics, Law and Economics*, 20(2), 255–269.

Chapter 3

1 United Nations (2022). Least Developed Countries (LDCs). Accessed on 5 September 2022 at [www.un.org/development/desa/dpad/least-developed-country-category.html].

2 Cambodianess (13 January 2020). *Value of Construction Projects Doubles in 2019*. By Ou Sokmean.

3 HM Government. *Modern Slavery*. Accessed on 15 February 2022 at [www.gov.uk].

4 Parsons, L., and Ly, V. L. (2020). *A Survey of the Cambodian Brick Industry: Population, Geography, Practice*. Phnom Penh: Brick Workers Trade Union of Cambodia.

5 Brickell, K., Parsons, L., Natarajan, N., and Chann, S. (2018). *Blood Bricks: Untold Stories of Modern Slavery and Climate Change from Cambodia*. Egham: Royal Holloway.

6 House of Commons (2021). *Investment Industry's Exposure to Modern Slavery*. Research Briefing Number CBP 9353. By Ali Shalchi, 25 October 2021. Accessed on 20 May 2022 at [http://commonslibrary.parliament.uk].

212

Notes

7   Observatory of Economic Complexity [OEC] (2019). *Bricks/Brick Trade.* Accessed on 29 November 2019 at [https://oec.world].

8   Ibid.

9   Parsons, L., Safra de Campos, R., Moncaster, A., Siddiqui, T., Cook, I., Abenayake, C., Jayasinghe, A., Mishra, P., Bilah, T., and Scungio, L. (2021). *Disaster Trade: The Hidden Footprint of UK Production Overseas.* Egham: Royal Holloway.

10 Her Majesty's Revenue and Customs [HMRC] (2020). *UK Brick Import Data.* Private Communication via the Brick Development Association.

11 Ibid.

12 Mishra, D. K. (2020). Seasonal migration and unfree labour in globalising India: insights from field surveys in Odisha. *Indian Journal of Labour Economics*, 63(4), 1087–1106.

13 BBC News (2 September 2014). *Why India's Brick Kiln Workers 'Live Like Slaves'.* By Humphrey Hawksley.

14 HM Government (2022). *Food Statistics in your Pocket: Prices and Expenditure.* Accessed on 10 February 2022 at [www.gov.uk/government/statistics].

15 The Economist (29 January 2020). *Interest in Veganism is Surging.*

16 Falguera, V., Aliguer, N., and Falguera, M. (2012). An integrated approach to current trends in food consumption: moving toward functional and organic products? *Food Control*, 26(2), 274–281.

17 Ibid.

18 Sexton, A. E., Garnett, T., and Lorimer, J. (2022). Vegan food geographies and the rise of Big Veganism. *Progress in Human Geography*, 46(2), 605–628.

19 The New York Times (30 September 1970). *F. T. C. Says Chevron's Claims For a Gas Additive Are False.*

20 Ibid.

21 Corpwatch (2001). *Greenwash Fact Sheet.* Accessed on 20 January 2021 at [www.corpwatch.org/article/greenwash-fact-sheet].

22 Beckert, S. (2014). *Empire of Cotton: A Global History.* New York: Penguin Random House USA.

23 Temin, P. (2013). *The Roman Market Economy.* Princeton: Princeton University Press.

24 Hansen, V. (2012). *The Silk Road.* Oxford: Oxford University Press.

25 Quiroga-Villamarín, D. R. (2020). Normalising global commerce: containerisation, materiality, and transnational regulation (1956–68). *London Review of International Law*, 8(3), 457–477.

Notes

26 Rodrigue, J. P. (2012). The geography of global supply chains: evidence from third-party logistics. *Journal of Supply Chain Management*, 48(3), 15–23.

27 Jedermann, R., Praeger, U., Geyer, M., and Lang, W. (2014). Remote quality monitoring in the banana chain. *Philosophical Transactions of the Royal Society A: Mathematical, Physical and Engineering Sciences*, 372(2017), 1–21.

28 Silvestre, B. S., Viana, F. L. E., and de Sousa Monteiro, M. (2020). Supply chain corruption practices circumventing sustainability standards: wolves in sheep's clothing. *International Journal of Operations and Production Management*, 40(12), 1873–1907.

29 Lawreniuk, S. (2020). Necrocapitalist networks: COVID-19 and the 'dark side' of economic geography. *Dialogues in Human Geography*, 10(2), 199–202.

30 Brickell, K., Lawreniuk, S., Chhom, T., Mony, R., So, H., and McCarthy, L. (2022). 'Worn out': debt discipline, hunger, and the gendered contingencies of the COVID-19 pandemic amongst Cambodian garment workers. *Social & Cultural Geography*, 1–20.

31 Obinger, H., and Schmitt, C. (2020). World war and welfare legislation in western countries. *Journal of European Social Policy*, 30(3), 261–274.

32 Amengual, M., and Distelhorst, G. (2020). *Cooperation and Punishment in Regulating Labor Standards: Evidence from the Gap Inc Supply Chain*. Available at SSRN 3466936: http://dx.doi.org/10.2139/ssrn.3466936.

33 Ibid.

34 Caro, F., Lane, L., and Saez de Tejada Cuenca, A. (2021). Can brands claim ignorance? Unauthorized subcontracting in apparel supply chains. *Management Science*, 67(4), 2010–2028.

35 Parsons, L., Safra de Campos, R., Moncaster, A., Siddiqui, T., Cook, I., Abenayake, C., Jayasinghe, A., Mishra, P., Bilah, T. and Scungio, L. (2021). *Disaster Trade: The Hidden Footprint of UK Production Overseas*. Egham: Royal Holloway.

36 Buller, A. (2022). *The Value of a Whale: On the Illusions of Green Capitalism*. Manchester: Manchester University Press.

37 Initiative Lieferkettengesetz (2022). *What the New SUPPLY CHAIN ACT Delivers – and What It Doesn't*. Accessed on 5 September 2022 at [https://lieferkettengesetz.de/].

Notes

Chapter 4

1 Kalandides, A., and Grésillon, B. (2021). The Ambiguities of 'Sustainable' Berlin. *Sustainability*, 13(4), 1666, 1–13.

2 Deloitte (2020). *Shifting Sands: The Changing Consumer Landscape*. London: Deloitte.

3 Praskievicz, S. (2021). How the environment became global. *Anthropocene*, 35, 100–305.

4 Cosgrove, D. (1994). Contested global visions: one-world, whole-earth, and the Apollo space photographs. *Annals of the Association of American Geographers*, 84(2), 270–294.

5 The Virginia Mercury (21 August 2019). *On Climate Change, We're All In This Together*. By Ivy Main.

6 Pope Francis (2015). *Laudato Si*. Accessed on 25 May 2022 at [www.vatican.va/].

7 The Independent (11 January 2021). *'Abusing Our Planet as if We Had a Spare One': UN Chief Leads Call for Action on Nature Crisis at Summit*. By Daisy Dunne.

8 Ritchie, H. (2020). *Where in the World do People Have the Highest CO2 Emissions from Flying?* Our World in Data. Accessed on 26 May 2022 at [https://ourworldindata.org/].

9 Ritchie, H., Roser, M., and Rosado, P. (2020). *CO₂ and Greenhouse Gas Emissions*. Our World in Data. Accessed on 26 May 2022 at [https://ourworldindata.org/].

10 National Oceanic and Atmospheric Administration [NOAA] (2022). *Trends in Atmospheric Carbon Dioxide*. Accessed on 26 May 2022 at [https://gml.noaa.gov/ccgg/trends/mlo.html].

11 co2levels.org (2021). *Atmospheric CO2 Levels Graph*. Accessed on 5 July 2021 at [http:// co2levels.org].

12 Climate Action Tracker (2021). *Climate Action Tracker*. Accessed on 5 July 2021 at [https://climateactiontracker.org/].

13 Committee on Climate Change [CCC] (2019). *Reducing UK Emissions 2019: Progress Report to Parliament*. London: CCC.

14 Climate Action Tracker (2021). *Climate Action Tracker*. Accessed on 5 July 2021 at [https://climateactiontracker.org/].

15 Baumert, N., Kander, A., Jiborn, M., Kulionis, V., and Nielsen, T. (2019). Global outsourcing of carbon emissions 1995–2009: a reassessment. *Environmental Science & Policy*, 92, 228–236.

Notes

16 Moran, D., Hasanbeigi, A., and Springer, C. (2018). *The Carbon Loophole in Climate Policy: Quantifying the Embodied Carbon in Traded Products.* San Francisco: KGM & Associates, Global Efficiency Intelligence, Climate Work Foundations.

17 Peters, G. P., Andrew, R. M., and Karstensen, J. (2016). *Global Environmental Footprints: A Guide to Estimating, Interpreting and Using Consumption-Based Accounts of Resource Use and Environmental Impacts.* Copenhagen: Nordic Council of Ministers.

18 Dehm, J. (2016). Carbon colonialism or climate justice: interrogating the international climate regime from a TWAIL perspective. *Windsor Yearbook of Access to Justice*, 33(3), 129–161.

19 Office of National Statistics [ONS] (2019). *The Decoupling of Economic Growth from Carbon Emissions: UK Evidence.* Accessed on 23 October 2019 at [www.ons.org.uk].

20 Ward, M. (2020). *Statistics on UK–EU Trade.* Briefing Paper Number 7851. London: House of Commons Library.

21 Department for Environment, Food and Rural Affairs [Defra] (2017). *The UK's Carbon Footprint 1997–2016.* Accessed on 2 December 2019 at [https://assets.publishing.service.gov.uk].

22 World Wildlife Fund (2020). *Carbon Footprint: Exploring the UK's Contribution to Climate Change.* London: WWF.

23 Intergovernmental Panel on Climate Change [IPCC] (2018). *Global Warming of 1.5° C: An IPCC Special Report on the Impacts of Global Warming of 1.5° C Above Pre-industrial Levels and Related Global Greenhouse Gas Emission Pathways, in the Context of Strengthening the Global Response to the Threat of Climate Change, Sustainable Development, and Efforts to Eradicate Poverty.* Geneva: IPCC.

24 Open Apparel Registry (2022). *The Open Apparel Registry.* Accessed on 5 September 2022 at [http://openapparel.org].

25 Nature Climate Change Editorial (2018). The price of fast fashion. *Nature Climate Change*, 8(1), 1.

26 World Bank (23 September 2019). *How Much Do Our Wardrobes Cost to the Environment? Feature Story.* Accessed on 5 July 2021 at [http://worldbank.org].

27 Nature Climate Change Editorial (2018). The price of fast fashion. *Nature Climate Change*, 8(1), 1.

28 Human Rights Watch [HRW] (2017). *Tracing the Thread: The Need for Supply Chain Transparency in the Garment and Footwear Industry.* New York: Human Rights Watch.

Notes

29 Zenz, A. (2020). *Coercive Labor in Xinjiang: Labor Transfer and the Mobilization of Ethnic Minorities to Pick Cotton.* Washington: Center for Global Policy.

30 The Independent (23 April 2020). *UK Government Urged to Ban Import of Chinese Cotton Made Using Uighur Muslim Forced Labour.* By Adam Withnall.

31 Parsons, L., Safra de Campos, R., Moncaster, A., Siddiqui, T., Cook, I., Abenayake, C., Jayasinghe, A., Mishra, P., Bilah, T., and Scungio, L. (2021). *Disaster Trade: The Hidden Footprint of UK Production Overseas.* Egham: Royal Holloway.

32 Ibid.

33 Reuters (2020). *UK to Big Brands: Do More to Avoid Forced Labour in China's Xinjiang.* By Kieran Guilbert.

34 BBC (9 December 2021). *China Committed Genocide Against Uyghurs, Independent Tribunal Rules.* By Joel Gunter.

35 Parsons, L., Safra de Campos, R., Moncaster, A., Siddiqui, T., Cook, I., Abenayake, C., Jayasinghe, A., Mishra, P., Bilah, T., and Scungio, L. (2021). *Disaster Trade: The Hidden Footprint of UK Production Overseas.* Egham: Royal Holloway.

36 Her Majesty's Revenue and Customs [HMRC] (2020). *UK Brick Import Data.* Private Communication via the Brick Development Association.

37 UN News (2008). *Half of Global Population will Live in Cities by End of this Year, Predicts UN.* Accessed on 5 July 2021 at [http://news.un.org].

38 World Green Building Council [WGBC] (2019). *Bringing Embodied Carbon Upfront: Coordinated Action for the Building and Construction Sector to Tackle Embodied Carbon.* London: WGBC.

39 World Green Building Council [WGBC] (2021). *Principle One: Protect Health and Wellbeing.* Accessed on 5 July 2021 at [https://worldgbc.org/].

40 Climate and Clean Air Coalition (2020). *Bricks: Mitigating Black Carbon and Other Pollutants from Brick Production.* Accessed on 5 July 2021 at [www.ccacoalition.org/].

41 United Nations Environment Programme [UNEP] (2020). *Sustainable Trade in Resources: Global Material Flows, Circularity and Trade.* Nairobi: UNEP.

42 Brown, D., Boyd, D. S., Brickell, K., Ives, C. D., Natarajan, N., and Parsons, L. (2021). Modern slavery, environmental degradation and

Notes

climate change: fisheries, field, forests and factories. *Environment and Planning E: Nature and Space*, 4(2), 191–207.

43 Turner and Townsend (2019). *International Construction Market Survey 2019*. London: Turner and Townsend.

44 World Freight Rates (2021). *Freight Calculator*. Accessed on 6 July 2021 at [https://worldfreightrates.com/].

45 Her Majesty's Revenue and Customs [HMRC] (2020). *UK Brick Import Data*. Private Communication via the Brick Development Association.

46 Department for Environment, Food and Rural Affairs [Defra] (2018). *Clean Air Strategy 2018*. Accessed on 5 July 2021 at [https://assets.publishingservice.gov.uk].

47 World Bank (2018). *Enhancing Opportunities for Clean and Resilient Growth in Urban Bangladesh: Country Environmental Analysis*. Washington: World Bank.

48 Eil, A., Li, J., Baral, P., and Saikawa, E. (2020). *Dirty Stacks, High Stakes: An Overview of Brick Sector in South Asia*. Washington: World Bank.

49 Becqué, R., Dubsky, E., Hamza-Goodacre, D., and Lewis, M. (2018). *Europe's Carbon Loophole*. San Francisco: Climate Works Foundation.

50 World Green Building Council [WGBC] (2019). *Bringing Embodied Carbon Upfront: Coordinated Action for the Building and Construction Sector to Tackle Embodied Carbon*. London: WGBC.

51 Ibid.

52 Moore, S. A., Rosenfeld, H., Nost, E., Vincent, K., and Roth, R. E. (2018). Undermining methodological nationalism: cosmopolitan analysis and visualization of the North American hazardous waste trade. *Environment and Planning A: Economy and Space*, 50(8), 1558–1579.

53 Hickel, J., Dorninger, C., Wieland, H., and Suwandi, I. (2022). Imperialist appropriation in the world economy: drain from the global South through unequal exchange, 1990–2015. *Global Environmental Change*, 73, 102467.

54 Batel, S., and Devine-Wright, P. (2017). Energy colonialism and the role of the global in local responses to new energy infrastructures in the UK: a critical and exploratory empirical analysis. *Antipode*, 49(1), 3–22.

55 Lyons, K., and Westoby, P. (2014). Carbon colonialism and the new land grab: plantation forestry in Uganda and its livelihood impacts. *Journal of Rural Studies*, 36, 13–21.

56 Alexander, C., and Stanley, A. (2021). The colonialism of carbon capture and storage in Alberta's Tar Sands. *Environment and Planning E: Nature and Space*, 5(4), 2112–2131.

Notes

57 Hoefle, S. W. (2013). Beyond carbon colonialism: frontier peasant livelihoods, spatial mobility and deforestation in the Brazilian Amazon. *Critique of Anthropology*, 33(2), 193–213.

58 Moran, D., Hasanbeigi, A., and Springer, C. (2018). *The Carbon Loophole in Climate Policy: Quantifying the Embodied Carbon in Traded Products*. San Francisco: KGM and Associates, Global Efficiency Intelligence, Climate Work Foundations.

59 Liboiron, M. (2021). *Pollution Is Colonialism*. Durham: Duke University Press.

60 Eurostat (2020). *Waste Shipment Statistics*. Accessed on 6 July 2021 at [https://ec.europa.eu/eurostat/].

61 Bachram, H. (2004). Climate fraud and carbon colonialism: the new trade in greenhouse gases. *Capitalism Nature Socialism*, 15(4), 5–20.

62 Alexander, C., and Stanley, A. (2021). The colonialism of carbon capture and storage in Alberta's Tar Sands. *Environment and Planning E: Nature and Space*, 25148486211052875.

63 Liboiron, M. (2021). *Pollution Is Colonialism*. Durham: Duke University Press.

64 Novotny, V., and Krenkel, P. A. (1975). A waste assimilative capacity model for a shallow, turbulent stream. *Water Research*, 9(2), 233–241.

65 Moran, D., Hasanbeigi, A., and Springer, C. (2018). *The Carbon Loophole in Climate Policy: Quantifying the Embodied Carbon in Traded Products*. San Francisco: KGM & Associates, Global Efficiency Intelligence, Climate Work Foundations.

66 Bhojvaid, V. (2021). Hazy clouds: making black carbon visible in climate science. *Journal of Material Culture*, 1–16. DOI: 10.1177/1359183521994864.

67 Harris, P. (2021). *Pathologies of Climate Governance: International Relations, National Politics and Human Nature*. Cambridge: Cambridge University Press.

68 Bhambra, G. K., Gebrial, D., and Nişancıoğlu, K. (2018). *Decolonising the University*. London: Pluto Press.

69 Simmonds, N. (2011). Mana wahine: decolonising politics. *Women's Studies Journal*, 25(2), 11–25.

70 Pahuja, S. (2011). *Decolonising International Law: Development, Economic Growth and the Politics of Universality*. Cambridge: Cambridge University Press.

Notes

71 Chao, S., and Enari, D. (2021). Decolonising climate change: a call for beyond-human imaginaries and knowledge generation. *eTropic: Electronic Journal of Studies in the Tropics*, 20(2), 32–54.

Chapter 5

1 Hinson, S., and Bolton, P. (2021). *Fuel Poverty*. House of Commons Library Research Briefing, 8 July 2021. Accessed on 16 January 2023 at [https://commonslibrary:parliament.uk/].

2 BBC (20 March 2016). *Fuel poverty: An Anatomy of a Cold Home*. By Datshiane Navanayagam.

3 Hinson, S., and Bolton, P. (2021). *Fuel Poverty*. House of Commons Library Research Briefing, 8 July 2021. Accessed on 16 January 2023 at [https://commonslibrary.parliament.uk/].

4 Science Norway (26 February 2016). *Humans are Tropical Animals*. By Georg Mathison.

5 World Bank (2021). *Climate Knowledge Portal: Cambodia*. Accessed on 24 May 2022 at [https://climateknowledgeportal.worldbank.org/].

6 Eyler, B., and Weatherby, C. (2019). *Letters from the Mekong: Toward A Sustainable Water-Energy Food Future in Cambodia*. Washington, DC: The Stimson Center.

7 Ibid.

8 Behringer, W. (2010). *A Cultural History of Climate*. Cambridge: Polity.

9 Solomon, S. (2007). *IPCC (2007): Climate Change The Physical Science Basis*. In American Geophysical Union Fall Meeting Abstracts (Vol. 2007), U43D-01.

10 Weber, E. U. (2010). What shapes perceptions of climate change? *Wiley Interdisciplinary Reviews: Climate Change*, 1(3), 332–342.

11 Parsons, L., and Nielsen, J. Ø. (2021). The subjective climate migrant: climate perceptions, their determinants, and relationship to migration in Cambodia. *Annals of the American Association of Geographers*, 111(4), 971–988.

12 Boyd, D. S., Jackson, B., Wardlaw, J., Foody, G. M., Marsh, S., and Bales, K. (2018). Slavery from space: demonstrating the role for satellite remote sensing to inform evidence-based action related to UN SDG number 8. *ISPRS Journal of Photogrammetry and Remote Sensing*, 142, 380–388.

Notes

13 Ayeb-Karlsson, S., Van der Geest, K., Ahmed, I., Huq, S., and Warner, K. (2016). A people-centred perspective on climate change, environmental stress, and livelihood resilience in Bangladesh. *Sustainability Science*, 11(4), 679–694.

14 Parsons, L., Safra de Campos, R., Moncaster, A., Siddiqui, T., Cook, I., Abenayake, C., Jayasinghe, A., Mishra, P., Bilah, T., and Scungio, L. (2021). *Disaster Trade: The Hidden Footprint of UK Production Overseas*. Egham: Royal Holloway.

Chapter 6

1 Shatkin, G. (1998). 'Fourth World' cities in the global economy: the case of Phnom Penh, Cambodia. *International Journal of Urban and Regional Research*, 22(3), 378–393.

2 Natarajan, N., Parsons, L., and Brickell, K. (2019). Debt-bonded brick kiln workers and their intent to return: towards a labour geography of smallholder farming persistence in Cambodia. *Antipode*, 51(5), 1581–1599.

3 Global Forest Watch (2022). *Cambodia*. Accessed on 30 May 2022 at [https://globalforestwatch.org].

4 Sultana, F. (2011). Suffering for water, suffering from water: emotional geographies of resource access, control and conflict. *Geoforum*, 42(2), 163–172.

5 Royal Government of Bhutan (2019). *National Strategic Development Plan (2019–2023)*. Thimpu: Royal Government of Bhutan.

6 World Bank (2022). *Bhutan Overview: Development News, Research, Data*. Accessed on 28 February 2022 at [https://worldbank.org].

7 World Bank South Asia (13 February 2020). Tweet.

8 Ritchie, H., and Roser, *M. Deforestation and Forest Loss: Which Countries are Gaining and Which are Losing Forest*. Our World in Data. Accessed on 28 April 2021 at [https://ourworldindata.org].

9 World Bank (2022). *Bhutan Overview: Development News, Research, Data*. Accessed on 28 February 2022 at [https://worldbank.org].

10 Katel, O. (2016). *Addressing Climate Change Concerns in Bhutan Himalaya*. Governance Today. Accessed on 16 January 2023 at [www.governancetoday.co.in/].

11 The Guardian (24 July 2014). *Put a Price on Nature? We Must Stop this Neoliberal Road to Ruin*. By George Monbiot.

Notes

12 Hardin, G. (1968). The tragedy of the commons: the population problem has no technical solution; it requires a fundamental extension in morality. *Science*, 162(3859), 1243–1248.

13 Cox, S. J. B. (1985). No tragedy of the commons. *Environmental Ethics*, 7(1), 49–61.

14 Wall, D. (2014). The commons in history: culture, conflict, and ecology. *International Journal of the Commons*, 9(1): 466–468.

15 The Phnom Penh Post (19 May 2015). *Tigers, Cobras Wished on Firm*. By Phak Seangly.

16 Kent, A. (9 November 2020). *The Desertion of Cambodia's Spirits*. New Mandala. Accessed on 5 April 2022 at [https://newmandala.org].

17 Bird-David, N. (1999). 'Animism' revisited: personhood, environment, and relational epistemology. *Current Anthropology*, 40(S1), S67–S91.

18 Borràs, S. (2016). New transitions from human rights to the environment to the rights of nature. *Transnational Environmental Law*, 5(1), 113–143.

19 Ibid.

20 Cullinane, C., and Montacute, R. (2018). *Pay as You Go? Internship Pay, Quality and Access in the Graduate Jobs Market*. London: The Sutton Trust.

21 Ibid.

22 The Sutton Trust (2019). *Elitist Britain: The Educational Backgrounds of Britain's Leading People*. London: The Sutton Trust.

23 Morgan, A., Clauset, A., Larremore, D., LaBerge, N., and Galesic, M. (2021). *Socioeconomic Roots of Academic Faculty*. Preprint. Accessed on 26 May 2022 at [https://doi.org/10.31235/osf.io/6wjxc].

24 Webometrics (2022). *Ranking Web of Universities*. Accessed on 26 May 2022 at [https://webometrics.org].

25 Collyer, F. M. (2018). Global patterns in the publishing of academic knowledge: Global North, global South. *Current Sociology*, 66(1), 56–73.

26 Nielsen, M. W., and Andersen, J. P. (2021). Global citation inequality is on the rise. *Proceedings of the National Academy of Sciences*, 118(7), e2012208118.

27 National Geographic (31 January 2020). *Southeast Asia's Most Critical River is Entering Uncharted Waters*. By Stefan Lovgren.

28 Hoang Thi Ha and Farah Nadine Seth (2020). The Mekong River ecosystem in crisis: ASEAN cannot be a bystander. *ISEAS Yusof Ishak Institute Perspective*, 2021(69), 1–9.

Notes

29 Seiff, A. (2022). *Troubling the Water: A Dying Lake and a Vanishing World in Cambodia*. Lincoln: University of Nebraska Press.

30 Eyler, B., and Weatherby, C. (2019). *Letters from the Mekong: Toward A Sustainable Water-Energy Food Future in Cambodia*. Washington, DC: The Stimson Center.

31 Ibid.

32 Gerlak, A. K., and Haefner, A. (2017). Riparianization of the Mekong River Commission. *Water International*, 42(7), 893–902.

33 Turnhout, E. (2018). The politics of environmental knowledge. *Conservation and Society*, 16(3), 363–371.

34 Pereira, M. D. M. (2017). *Power, Knowledge and Feminist Scholarship: An Ethnography of Academia*. London: Taylor & Francis.

35 Boisselle, L. N. (2016). Decolonizing science and science education in a postcolonial space (Trinidad, a developing Caribbean nation, illustrates). *Sage Open*, 6(1), 2158244016635257.

36 Lotz-Sisitka, H. B. (2017). Decolonising as future frame for environment and sustainability education. In Corcoran, P., and Weakland, J. (eds). *Envisioning Futures for Environment and Sustainability Education*. Wageningen: Wageningen Academic Publishers, 45–62.

37 Merchant, C. (1980). *The Death of Nature: Women, Ecology, and the Scientific Revolution*. London: Harper Collins.

38 Fanon, F. (1970: 18). *Black Skin, White Masks*. London: Pluto Press.

39 Fanon, F., Sartre, J. P., and Farrington, C. (1963: 188). *The Wretched of the Earth* (Vol. 36). New York: Grove Press.

40 Wa Thiong'o, N. (1992). *Decolonising the Mind: The Politics of Language in African Literature*. Nairobi: East African Publishers.

41 Higham, C., and Kerry, H. (2022). *Taking Companies to Court over Climate Change: Who is Being Targeted?* LSE Commentary. Accessed on 6 September 2022 at [www.lse.ac.uk/granthaminstitute].

42 Ibid.

Chapter 7

1 Nordhaus, W. D. (1991). To slow or not to slow: the economics of the greenhouse effect. *Economic Journal*, 101(407), 920–937; Ekins, P. (1999). *Economic Growth and Environmental Stewardship: The Prospects for Green Growth*. London: Routledge.

Notes

2 Hickel, J., and Kallis, G. (2020). Is green growth possible? *New Political Economy*, 25(4), 469–486.

3 Dittrich, M., Giljum, S., Lutter, S., and Polzin, C. (2012). *Green Economies around the World: Implications of Resource Use for Development and the Environment*. Vienna: Sustainable Europe Research Institute.

4 Hickel, J. (2020). *Less Is More: How Degrowth Will Save the World*. London: Random House.

5 Thunberg, G. (2019). *No One Is Too Small to Make a Difference*. London: Penguin.

6 Wallace, J. (17 March 2011). *In Cambodia, Gambling on the Rain*. The Atlantic.

7 The Financial Times (3 June 2016). *Britain Has Had Enough of Experts, Says Gove*. By Henry Mance.

8 Trump, D. J. (6 November 2012). Tweet.

9 Edwards, P. N. (2010). *A Vast Machine: Computer Models, Climate Data, and the Politics of Global Warming*. Cambridge: MIT Press.

10 Demeritt, D. (2001). The construction of global warming and the politics of science. *Annals of the Association of American Geographers*, 91(2), 307–337.

11 Pincebourde, S., Murdock, C. C., Vickers, M., and Sears, M. W. (2016). Fine-scale microclimatic variation can shape the responses of organisms to global change in both natural and urban environments. *Integrative and Comparative Biology*, 56(1), 45–61.

12 Staal, A., Flores, B. M., Aguiar, A. P. D., Bosmans, J. H., Fetzer, I., and Tuinenburg, O. A. (2020). Feedback between drought and deforestation in the Amazon. *Environmental Research Letters*, 15(4), 1–9.

13 Hulme, M. (2009). *Why we Disagree about Climate Change: Understanding Controversy, Inaction and Opportunity*. Cambridge: Cambridge University Press.

14 Ward, B. (2012). *Desperate Shenanigans as Climate Change 'Sceptics' Try to Misrepresent IPCC Report*. Grantham Institute Commentary. Accessed on 17 January 2023 at [www.lse.ac.uk/granthaminstitute/news/desperate-shenanigans-as-climate-change-sceptics-try-to-misrepresent-ipcc-report/].

15 Chow, W., Dawson, R., Glavovic, B., Haasnoot, M., Pelling, M., and Solecki, W. (2022). *IPCC Sixth Assessment Report (AR6): Climate Change 2022-Impacts, Adaptation and Vulnerability: Factsheet Human Settlements*. Geneva: Intergovernmental Panel on Climate Change; Allen, M.,

Notes

Babiker, M., Chen, Y., and de Coninck, H. C. (2018). *IPCC SR15: Summary for Policymakers. In IPCC Special Report Global Warming of 1.5 °C.* Geneva: Intergovernmental Panel on Climate Change.

16 Smith, J., Tyszczuk, R., and Butler, R. (2014). *Culture and Climate Change: Narratives (Vol. 2).* Cambridge: Shed.

17 Mann, M. E. (2021). *The New Climate War: The Fight to Take Back our Planet.* London: Hachette UK.

18 Shell (2022). *What Sustainability Means at Shell.* Accessed on 6 September 2022 at [www.shell.com/sustainability/our-approach/sustainability-at-shell.html].

19 BP (2022). *Reimagining Energy for People and our Planet.* Accessed on 6 September 2022 at [www.bp.com/].

20 Gilbert, K. (2021). *Are Governments at COP26 Guilty of 'Greenwashing?'* Columbia Business School. Accessed on 30 September 2022 at [https://leading.gsb.columbia.edu/].

21 The National (17 February 2022). *UK Government 'Greenwashing to Extreme' over North Sea Oil and Gas Production.* By Abbi Garton-Crosbie.

22 Mongabay (15 March 2022) *Amazon Deforestation Starts 2022 on the Fastest Pace in 14 Years.* By Mongabay.

23 Greenpeace (19 November 2021). *Amazon Deforestation Rate 22% Higher Than Last Year.* By Katie Nelson.

Chapter 8

1 Summers, L. (1991). *The Memo.* World Bank Office of the Chief Economist. Accessed on 1 December 2008 at [www. whirledbank. org/ourwords/summers.html].

2 Baumeister, R. F., and Juola Exline, J. (1999). Virtue, personality, and social relations: self-control as the moral muscle. *Journal of Personality,* 67(6), 1165–1194.

3 Fein, H. (1979). *Accounting for Genocide: National Responses and Jewish Victimization during the Holocaust.* New York: Free Press.

4 Sultana, F. (2007). Reflexivity, positionality and participatory ethics: negotiating fieldwork dilemmas in international research. *ACME: An International Journal for Critical Geographies,* 6(3), 374–385.

5 Greenpeace UK (2021). *Trashed: How the UK is Still Dumping Plastic Waste on the Rest of the World.* London: Greenpeace UK.

Notes

6 Deloitte (2022). *Shifting Sands: Are Consumers Still Embracing Sustainability? Changes and Key Findings in Sustainability and Consumer Behaviour in 2021.* Accessed on 27 May 2022 at [https://deloitte.com].

7 Kantar (2021). Our planet issue: accelerating behaviour change for a sustainable future. *Public*, 4(October).

8 Kelman, I. (2020). *Disaster by Choice: How our Actions Turn Natural Hazards into Catastrophes.* Oxford: Oxford University Press.

9 Munasinghe, M., Deraniyagala, Y., Dassanayake, N., and Karunarathna, H. (2017). Economic, social and environmental impacts and overall sustainability of the tea sector in Sri Lanka. *Sustainable Production and Consumption*, 12, 155–169.

10 Berners-Lee, M. (2020). *How Bad are Bananas? The Carbon Footprint of Everything.* London: Profile Books.

11 Senanyake, K. (1993). Causes and mechanism of landslides in Sri Lanka. *International Conference on Environmental Management, Geo-Water and Engineering Aspects*, 323–326.

12 UNISDR (2019). *DesInventar Database.* Accessed on 1 February 2021 at [www.desinventar.lk].

13 Ibid.

14 Yuan, J., Lu, Y., Wang, C., Cao, X., Chen, C., Cui, H., ... and Du, D. (2020). Ecology of industrial pollution in China. *Ecosystem Health and Sustainability*, 6(1), 1779010.

15 Ntiamoah, A., and Afrane, G. (2008). Environmental impacts of cocoa production and processing in Ghana: life cycle assessment approach. *Journal of Cleaner Production*, 16(16), 1735–1740.

16 Flysjö, A. (2011). Potential for improving the carbon footprint of butter and blend products. *Journal of Dairy Science*, 94(12), 5833–5841.

17 Bananalink.org (2022). *The Problem with Bananas.* Accessed on 27 May 2022 at [www.bananalink.org.uk/].

18 Kantar (2021). Our planet issue: accelerating behaviour change for a sustainable future. *Public*, 4(October).

19 Mark Dolan, GB News, 3 April 2022.

20 Dyke, J., Watson, R., and Knorr, W. (22 April 2021). *Climate Scientists: Concept of Net Zero is a Dangerous Trap.* The Conversation.

21 The Ecologist (12 October 2015). *For Climate Change Action, we Must Fight Back against the Media Billionaires.* By Donnachadh McCarthy.

22 Parsons, L. (2021). Climate migration and the UK. *Journal of the British Academy*, 9, 3–26.

Notes

23 Boas, I. (2021). *Climate Migration Myths*. Presentation as part of panel 'The Battle for the Borders of Climate Science: Agnotology, Epistemology and the Contested Politics of Environmental Ignorance', at RGS-IBG Annual Conference, September 2021.

24 Nag, P. K., Ashtekar, S. P., Nag, A., Kothari, D., Bandyopadhyay, P., and Desai, H. (1997). Human heat tolerance in simulated environments. *Indian Journal of Medical Research*, 105, 226–234.

25 The Guardian (18 August 2022). *The Century of Climate Migration: Why we Need to Plan for the Great Upheaval*. By Gaia Vince.

26 De Bruyn, B. (2020). The great displacement: reading migration fiction at the end of the world. *Humanities*, 9(1), 1–16.

27 National Oceanic and Aerospace Administration (2022). *Global Monitoring Laboratory: Trends in Atmospheric Carbon Dioxide*. Accessed on 27 May 2022 at [https://gml.noaa.gov/].

28 Skuce, A. G., Cook, J., Richardson, M., Winkler, B., Rice, K., Green, S. A., Jacobs, P., and Nuccitelli, D. (2016). Does it matter if the consensus on anthropogenic global warming is 97% or 99.99%? *Bulletin of Science, Technology & Society*, 36(3), 150–156.

Index

abuse
  ending of 74
  of people and their environment 28
academic research 136–139
accountability 87
  see also carbon accounting
action on the environment 183
activism 170–171
agencies, environmental 10
Amazon basin 94
Angkor Wat temple 126
animal population 26
animal welfare 58
animist thinking 133
anthropology 133
anxiety 182
Apollo space programme 78
apprenticeship 42
Arkwright, Richard 41
assimilative capacity limits 95
Attenborough, David 190
attitudes of consumers 58
Australia 191

Bacon, Francis 167
bananas 189
Bangkok 13
Bangladesh 6–7, 36, 118–122, 153–154
Barclay, Frederick 193–194
Battambang 158–159, 164–165, 172
batteries 187–188
Beckert, Sven 40–41
Berlin 76, 88, 92
Better Cotton Initiative 85
Bhutan 127–130
Biden, Joe 173–174
biodiversity 45, 140
Black Rock (company) 54
Bolivia 134
Bolsonaro, Jair 174
bonded labour 53–56, 119
Borrás, Susana 133
BP (company) 173
brands 11, 69
Brazil 174
Bregman, Rutger 190
Brexit 161

228

Index

brick industry 2, 7, 28, 48–57, 88–92, 96, 114–122, 125–126, 148, 203
brick names 57
Britain *see* United Kingdom
British Broadcasting Corporation (BBC) 56

Cambodia 10–14, 29, 32–40, 43, 45, 48–53, 65–66, 71–72, 85, 87, 102, 106–111, 125–128, 132, 138, 142, 158–159, 162–163, 169, 178
  National Bank of 166
Canadia Bank tower 147
capital flight 11
capitalism 30, 170, 186
  green 73
carbon, hidden sources of 88
carbon accounting 81–83, 87, 96–97
carbon colonialism 6, 16, 20, 47, 82, 94–97, 205, 207
  use of the term 94
carbon costs 86, 96
carbon dioxide (CO2) 79–81, 89
carbon footprints 79, 84–87, 90, 128
carbon growth rate 199–200
carbon-intensive production 5, 8
carbon leakage 92
carbon sinks 128
Carney, Mark 154
Chevron (company) 17, 60–61
child labour 42, 52–56, 119
China 14, 65, 81, 84–86, 139, 142, 146, 188
chocolate biscuits 188
Citarum river, Java 3–4

climate
  in crisis 122–123, 182, 203
  as a cultural concept 111
  politics of 165–167, 173
  use of the term 112
climate change 2–8, 16–22, 39, 44, 47, 74, 76–79, 83–84, 87–92, 97–98, 103–104, 108–123, 140, 149, 154–156, 168–172, 190, 195, 207
  action on 21, 171
  construction of 88–92
  denial of 22, 194
  disagreement about 154
  'green growth' solution to 155–156
  human impacts of 6, 168
  human origins of 22
  perceptions of 113–114
  precarity of 114–123
  in scientific sense 116–117
  speaking about 179
  successful response to 98
  see also impact of climate change; vulnerability to climate change
Climate Change conference (Glasgow, 2021) 152–154
  see also COP26
climate consensus 16
climate-linked displacement 196
climate policy 97, 158, 169, 172–173
climate scepticism 172
climate science 201–202
climatology 164
clothing labels 1
cobalt mining 27–28
Coca Cola 60–62

Index

cognitive dissonance 7
cold, experience of 104–105
collective action 79
Collyer, Fran 139
colonialism 15, 28–29, 94–95, 186–187
  legacy of 186
  see also carbon colonialism
common land and common ownership 43, 130–131
complaints, blocking of 11
Congo, Democratic Republic of 28
constitutions 134
consumers' knowledge and power 19, 58–59, 62
consumption, scaling back of 156
Convergence Theory 36
'COP' 21, 26, 157, 169–170, 174, 199–200
Corbyn, Jeremy 88
Corden, James 78
corporate governance and self-governance 17, 74
corporate social responsibility 69–71
corruption 6
cost, environmental 6
Cotta, Benedetta 47
cotton production 41–42, 63, 85
Covid-19 pandemic 69, 170
Cox, Susan 130–131
cultural environment 131
Cyclone Nargis (2008) 16

Dachaoshan dam 141
Daily Mail 193
damage, environmental 2, 7–11, 15–20, 26, 46–47, 83, 108, 121–122, 201
  hiding of 93–94

dams 141–146
decolonisation 98
deforestation 26, 29, 72, 127, 174
  slash-and-burn type 94
'degrowth' 156
delayed gratification 178
Deloitte (company) 182
Demeritt, David 164
Department of Farming and Rural Affairs (Defra) 91
development, theories of 37
'displacement' 92
Dolan, Mark 192–193
drones 72
droughts 146
dumps 203
dung, trade in 32, 38–39
duty of care to workers 21

'Earthrise' photograph from the Moon 78
economic development 4
'Economic Development with Unlimited Supplies of Labor' 37
economies of scale 28
ecosystems 135, 155
Ecuador 134
El Niño 142
emissions 4–5, 8, 11, 17–20, 80–97, 122–123, 187, 199–202, 206
  human-made 201
enclosures 43
Energy Performance of Buildings Directive (EU, 2002) 91–92
enforcement of standards 11, 52, 96
  see also self-enforcement

230

Index

English language 14
environment
    as an abstract commodity 148
    use of the word 77–78
environmental accountancy 94–95
environmental damage *see* damage
environmental history 46
environmental justice 123
environmental pressures 117–121
environmental security 22
environmentalism 21–22, 60,
    170–171
    contemporary problem in 127
    success of 62
ethics 84–86
    of consumption 59
European Union (EU) 20, 81–82,
    91, 94
everyday challenges and needs
    12, 190
everyday life and labour 8, 117
everyday luxuries 126
experience
    of climate breakdown 205
    gaining of 137
exploitation 131, 141
    of labour 28, 44
exports
    of the climate crisis 203
    environmentally regulated 82
Extinction Rebellion 4, 170
extraction of resources 27, 29,
    155–156
extractivism 149

Fanon, Frantz 149
farming methods 32, 38–40
Federal Trade Commission 61
Fein, Helen 179

feminism 168
fertilisers 46, 115
fieldwork 48–49
fisheries 46, 107–110, 113
flight paths and flight times 12
flooding 5–7, 68, 129
flying 79–80
food industry 58–59
Francis, Pope 79
'Fridays for Future' 4, 154
fuel poverty 104

Gap (company) 70
garment industry 34–37, 41–45,
    51, 67–68, 71, 84–88, 95
Gates, Bill 154
geography
    of climate change 22
    'death' of 12
geopolitical shifts 31
German language 111
Germany 73, 84, 93
Ghana 188
Glasgow 152, 154, 157–158
Glasgow Landing 170
global factory, the, lost knowledge
    in 65–75
global warming 16, 89, 96, 161
    human-caused 16
globalisation 10–12, 39, 41,
    45–46, 63–64, 94, 97, 135,
    148, 180, 184
    history of 19
Good Housekeeping 187
Gore, Al 201
governance, environmental 149
Les Grands Lacs de Seine 5
Greece 30
green-growth modelling 17

Index

green policy 174
  see also climate policy
greenhouse gases 16, 89
Greenpeace 174
'greenwashing' 2, 17, 19, 61–62, 73, 154, 173–174, 182, 207
Greenworks (cleaning fluid) 61
gross domestic product (GDP) 155–156
  decoupling from resource use 155–156
growth, economic 22, 29, 45–46, 155–156
Gualinga, Nina 149–150
Guterres, Antonio 79, 173–174

Hardin, Garett 129–131
heat, extreme 65
hemispheres, Northern and Southern 94
Hickel, Jason 156
hidden costs and impacts 8, 127
Himalayas 128
history, knowledge of 93
human activities 84, 201–202
  remoulding the natural world 146
human cost of environmental damage 28
human qualities in the natural environment 133
Humboldt University 92
Hurricane Katrina (2005) 16

ignorance, lucrative 15
impact of climate change 110, 113, 121, 150, 189, 201, 207
imperialism 47, 93
India 88, 139

Indonesia 9
industrial revolution 26
industrial workforce, creation of 31–40, 43
industrialisation 30–31, 36, 39–40, 44, 63
inequality, economic 122–123
inspection of industrial processes 11–12, 15, 67
Intergovernmental Panel on Climate Change (IPCC) 111–112, 167–168, 201
international agreements 80
international law 134
International Maritime Organisation 157
internships 137–138
Islamabad 196

Jakarta 6–7
'Jekyll and Hyde' development 146
Johnson, Boris 199

Kampot 29
Kentucky Fried Chicken (KFC) 59
Khmer Rouge 33, 114, 135
knowledge
  power of 140
  creation and control of 145
  infrastructure of 148
Kuznets Curve, Environmental 4

labour law 43
land
  acquisition of 132
  sales of 125–126
  use of 43
landlessness 43

232

Index

scientific revolution 45
self-enforcement 83
7-Up 61
Sierra Club v. Morton, case of (1972) 134
Silk Road 63
Singapore 30
slavery 41
  modern 53–54, 89
small actions by individuals 78
smallholders 43–44, 118, 142
Social Attitudes Survey, UK 171
social narratives 179
social science 168–169
Sorya Mall 125
sourcing of materials 85
South Korea 30
speaking for the environment 135
species extinction 26
Sri Lanka 185–186
Standard Life (company) 54
standards, environmental and for working conditions 9, 11, 69–70
Stockholm Declaration (1972) 133–134
subcontracting 13, 70
Summers, Lawrence 176, 180
supply chains 8–20, 53, 56, 59, 62, 66–74, 83–87, 90, 95–98, 207
  accountability in 65, 68
  global 206
  international 59, 65, 67
sustainability 7–21, 44, 46, 70, 74–77, 84, 87, 89, 130–131, 149, 155–157, 170, 175, 187, 190
  aesthetics of 181

at home 198–201
commitment to 62
narrative on 183

tax 190
tea plantations 185
teabags 184–185
temperature 83, 195, 202
textile industry 40–47
Thames Barrier 5
Thiong'o, Ngugi wa 150
Thunberg, Greta 157, 174
tipping point 130
toasters 8–9
Tonle Sap Lake and Tonle Sap Authority (TSA) 29, 106–110, 113, 117
toxic waste 6, 176
trading relations, inequality of 92
traditional way of life 42–43
tragedy of the commons 130, 133
transitional stages 30–31, 40, 44, 124–125
transparency, lack of 87
travel on business 13
Trump, Donald 161
truth as distinct from partial results 168
Tu brand 13–14

Uyghurs 86
United Arab Emirates (UAE) 79
United Kingdom 20, 41, 45, 77–90, 138, 161, 174, 180
United Nations 51–52, 162, 183, 199–200
  Declaration on the Human Environment (1972) 133–134

Index

United Nations (*Cont.*)
  Framework Convention on Climate Change (UNFCCC) 80–81
  High Commissioner on Human Rights 51
  Human Rights Council 134
United States 58–61, 79, 81, 86, 138–139
urban development 37
urban living 88–89
urbanisation 121

veganism 58–59
Venice 5
Vietnam 85
Virginia Mercury (newspaper) 78–79
vulnerabiity to climate change 20, 119, 122, 184–187, 205–206

wage labour 114–115
Wall, Derek 131

waste 15–16, 47, 176
  see also toxic waste
wastewater 188
water level 197
water shortages 142
weather 165
Westervelt, Jay 60
working conditions 42, 56, 205
working environment 119
World Bank 128, 130, 133, 176, 180
World Climate Conference (2016) 80–81
World Economic Forum (Davos) 190
World Meteorological Association 165
Wutty, Chut 171

Xinjiang 85

Yay Mom 101, 114, 117–118, 122
Yousof Ishak Institute 140